Superconductors
Materials and Applications

Edited by

Inamuddin[1], Tariq Altalhi[2], Vikas Gupta[3], Mohammad Luqman[4]

[1]Department of Applied Chemistry, Zakir Husain College of Engineering and Technology, Faculty of Engineering and Technology, Aligarh Muslim University, Aligarh-202002, India

[2]Department of Chemistry, College of Science, Taif University, 21944 Taif, Saudi Arabia

[3]Department of Chemistry, Motherhood University, Roorkee-Dehradun Road, Vill. Karoundi Post Bhagwanpur, Tehsil - Roorkee, Distt. Haridwar, Uttrakhand-247661, India

[4]Department of Chemical Engineering, College of Engineering, Taibah University, Yanbu, Saudi Arabia

Published by **Materials Research Forum LLC**
Millersville, PA 17551, USA

Published as part of the book series
Materials Research Foundations
Volume 132 (2022)
ISSN 2471-8890 (Print)
ISSN 2471-8904 (Online)

Print ISBN 978-1-64490-210-3
eBook ISBN 978-1-64490-211-0

Distributed worldwide by

Materials Research Forum LLC
105 Springdale Lane
Millersville, PA 17551
USA
https://www.mrforum.com

Manufactured in the United States of America
10 9 8 7 6 5 4 3 2 1

Table of Contents

Preface

A superconductor is a material that conducts electricity to flow current impeded without electrical resistance and expel magnetic flux fields. Since the discovery of superconductors in 1911, materials with superconductivity took the place as frontiers in commercial applications. Classification of superconductors is based on material properties such as critical temperature, theory of operation and magnetic fields. Superconductors connect theoretical physics, engineering and practical applications. Superconductors are utilized in power transmission, particle accelerators, transportation, equipment, electromagnets, computing, environment, electronics, medical, defense and large-scale applications. There is, therefore, a necessity to grasp the knowledge of superconductors applications in modern society.

The book **Superconductors - Materials and Applications** aims to discuss the design, properties and applications of superconductor materials. Topics include various categories of superconductors such as type-I, type-II, bulk, hard, soft, oxide, fermions, organic, iron, Lanthanide-based superconductors and so on. The book also focuses on applications of superconductor technology in interdisciplinary applications. The book is written by keynoted experts and scientists in this field. It is an archival reference for applied physics, material scientists, engineers, and policymakers in science and technology. This book contains 13 chapters and chapters are summarized as given below:

Chapter 1 highlights the historical background, theoretical knowledge as well as the difference between perfect conductors and superconductors. It describes how a normal conductor behaves as a superconductor at critical values of temperature and magnetic field. Moreover, it also covers the types and basic fundamental properties of superconductors in detail.

Chapter 2 deals with the fundamental properties of superconductors followed by the major types which are currently being investigated by several research groups. The basic idea of the Meissner effect is linked with superconductors. Some underlying concepts like the conduction of electrons and surface energy are discussed. Superconductor's types which are practically being implemented in devices such as Cuprates and iron-based superconductors are described in detail.

Chapter 3 comprises an introduction to superconductors and their fundamental characteristics. Types of superconductors such as type I and Type II superconductors, high Tc and low Tc superconductors, and organic and magnetic superconductors are briefly discussed. The basic characteristics and requirements of superconductors are deliberated.

Chapter 4 discusses the advanced applications of superconductors in the research field as well as in the industrial field. Most of the applications rely on LTS materials, but commonly HTS wires are used for economic advantages. Thus, superconductivity plays a vital role in the advancement of the human lifestyle.

Chapter 5 highlights the advances made toward the formation of lanthanide-based superconductors. Details about their classification, preparation methods and prospects in advanced technological applications are provided. Understanding their properties and characterization methods is very crucial in facilitating the synthesis of lanthanide-based superconductors with desirable superconducting properties for various applications.

Chapter 6 details various Type-I superconductors, their critical temperatures, history of superconductivity, physical properties, applications, recent research on Type-I superconductors, various theories proposed for explaining their properties, and different applications of superconductors.

Chapter 7 details various types of high-temperature bulk superconductors and their physical properties, applications, and recent research in the field.

Chapter 8 focused on Type 1 superconductors (soft superconductors) properties crystallographic structure and usage areas. This chapter also gives information about Type 2 superconductors and their differences from Type 1. In addition, the concept of superconductivity, the history of the process, its economy, and future expectations are mentioned in the chapter.

Chapter 9 aims to discuss different types (type 1 and type 2) of superconducting materials including their behavior towards the magnetic field. It discusses different existing theories and hypotheses on superconducting materials with emphasis on cooper pair theory. Applications of superconducting material in medical and geological fields have also been discussed.

Chapter 10 discusses the underlying physics of high-temperature superconductors (HTSCs) and the limitations of BCS theory in its explanation. HTSCs have a layered structure of Cu_2O which is distinguished by insulating layers. In this chapter Nickel and Bismuth based HTSCs are discussed followed by their potential application in fusion reactors, energy storage, and co-axial magnetic gears.

Chapter 11 entails the superconducting metamaterials which have considerably low losses concerning metamaterials. Types of superconducting metamaterials such as slow loss metamaterials and scaling of their properties are discussed. The scaling of split ring resonators is deliberated briefly. Other novel superconducting metamaterials such as superconducting composites and DC metamaterials are discussed.

Chapter 12 deals with the commercial superconductors that are being currently used in several medical applications. A considerable part of the annual GDP of several advanced countries goes into medical treatments and therefore, the prior diagnosis is quite important to save the said amount. Medical applications including magnetic resonance imaging, magnetic gene transfer, magnetic drug delivery system and cancer & internal hemorrhage detection which are mainly based on superconductors have been discussed in this chapter.

Chapter 13 deals with the superconductors developed over the years for their implementation in Magnetic Resonance Imaging (MRI) systems. MRI is a well-known and established diagnosis tool in medical procedures nowadays. Several superconductors including Nb-Ti and Nb_3Sn, copper-based superconductors, rare-earth barium copper oxide superconductors (REBCO), MgB_2 superconductors, and iron-based superconductors (IBS) are discussed.

Superconductors - Materials and Applications
Materials Research Foundations 132 (2022) 1- 16

Materials Research Forum LLC
https://doi.org/10.21741/9781644902110-1

Chapter 1

Basic Concepts and Properties of Superconductors

Shoomaila Latif*[1], Muhammad Husnain[1], M. Hassan Siddique[2], Muhammad Imran[3]

[1]School of Physical Sciences, University of the Punjab, Lahore-Pakistan

[2]Hi Tech Blending Pvt Ltd - ZIC Motor Oil, Lahore-Pakistan

[3]Centre for Inorganic Chemistry, School of Chemistry, University of the Punjab, Lahore-Pakistan

*shoomaila.sps@pu.edu.pk

Abstract

The phenomenon of super conductance is quite fascinating due to enormous applications and hence intense research in this area attracted engineers, scientists and businessmen. In this Chapter, we will briefly elaborate the conversion of a normal conductor to a superconductor which is a fascinating material since its discovery as well as the role of critical temperature and critical magnetic field for the super phenomenon of superconductivity. A short historical journey of superconductors from 1911 to date is also the part of this chapter that started with the work of Onnes on extreme low temperatures in cryogenic laboratories. The difference between perfect conductor and superconductor, classification of superconductors and finally the fundamental properties of superconductors have been discussed precisely.

Keywords

Superconductors, Critical Temperature, Critical Magnetic Effect, Meissner Effect

Contents

Materials Research Forum LLC
https://doi.org/10.21741/9781644902110-1

1.   Introduction and background

Conductors are the materials that let the electricity as well as the heat to pass through them for example metals, alloys, earth, animals and human body etc. But conductance of a material has a direct connection with the resistance of that material as resistance is one of the basic properties of the conductors. The resistance gives us qualitative as well as quantitative information about the flow of the electric current. Actually, electrical resistivity is the inverse of electrical conductivity. The conductors experience resistivity in the smooth flow of current at temperatures above the absolute zero [1].

This resistivity is because of the displacement of atoms from their equilibrium positions as a result of vibration motion of atoms at elevated temperatures. The freely moving conduction electrons carry the current through the conducting material and feel no resistivity in a crystalline structure with the atoms lying on a regular repetitive crystal lattice because electrons have a wave-like nature and it can pass through a perfectly periodic structure without being scattered in different directions. But the thermal vibrations of atoms at high temperature disrupt the regular symmetry of atoms causing a resistivity in the smooth flow of conduction electrons. However, on cooling the materials, resistivity decreases too. For pure metals, resistivity is almost zero at low temperatures but normally no material is perfectly pure [2]. The impurities in the materials cause resistivity even at low temperatures. Though, an unusual behaviour has been observed in certain metals on approaching the temperature a few degrees above the absolute zero. At this point the electrical resistance becomes almost zero and this is called the superconducting state of the materials [3].

So, the materials which have zero resistivity when they are subjected to a certain characteristic temperature are known as superconductors. The maximum temperature that is the characteristic of each material at which that material acts as a superconductor, is known as the **"critical temperature"** and denoted by **Tc**. And the maximum magnetic field at which a material behaves as superconductors, is called the **"critical field"** denoted as **Bc**. The magnitude of external magnetic fields larger than a critical value, terminate superconductivity. In fact, superconducting material cannot support a magnetic field inside it [4].

Theoretically, there are certain materials which exhibit this unusual behaviour of superconductivity. In point of fact, the superconductors are the normal conductors above the critical temperature [5]. The main difference between a superconductor and a normal conductor is that the normal conductor has finite resistance, on contrary to the superconductor which has zero electrical resistance. However, it is quite difficult to reach this superconducting state of the material which exists below the critical temperature. But once this state has been achieved by the material, the resistivity of the material falls to zero [6]. Recently, super conductors with high critical temperatures (130k) have been discovered.

Cuprate superconductors are considered as the main class whereas iron-based compounds as the second major class of high temperature superconductors [7]. To date, no material is known that can behave as a superconductor at ordinary conditions of temperature and pressure [8,9].

A cooling system is must for all high-temperature superconductors like liquid helium type problematic coolants (for metallic high temperature superconductors), liquid nitrogen type relatively friendly coolant (that can be used for the cooling of high temperature ceramic superconductors) but dry ice cannot work as coolant for superconductors [10].

2. History of superconductors

- A hundred years ago Heike Kamerlingh Onnes was the very first scientist whose work gave a new perspective to the researchers which became the basic cause of the marvellous phenomenon of super-conductance. He was working on the properties of materials at low temperatures in his cryogenic laboratory as shown in fig. 1. On his big achievement of liquefaction of helium, he was awarded the Nobel Prize in 1913 but during his work, he observed the sudden disappearance of resistivity mercury at the temperature of 4.2K as shown in fig. 2. This zero resistivity at low temperatures led to the great discovery of superconducting materials that was a breakthrough in the field of conductance at that time [11].

- Then his student Gilles Holst with the collaboration of Onnes measured the resistivity of mercury at low temperature. His work is shown in fig.2 [12].

- In 1933, W. Meissner and R. Ochsenfeld explained the exclusion of magnetic fields from inside the superconductors (now known as the Meissner effect) but this disappearance occurred under a definite critical field strength which depends on the critical temperature and the material of superconductors under study. Moreover, they revealed that superconductors are superior to perfect conductors in the field of electronics [13].

Fig.1: Kamerling Onnes at work in his laboratory

- In 1972, John Bardeen, Leon Neil Cooper and John Robert Schrieffer mutually evolved a theory of superconductivity i.e., known as the *BCS-theory* and for this collectively developed BCS theory, all the three US physicists were awarded the Nobel Prize in Physics. The three scientists have also put forward the quantum theory for superconductors in 1957 [14].

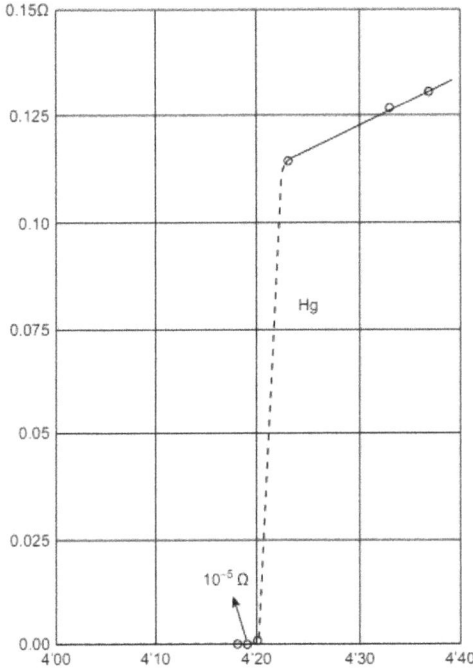

Fig.2: Resistivity of mercury at low temperature by Kamerlingh Onnes (1911)

- Another Noble prize was received by Brian Josephson in 1973 for a new improvement in the field of superconductors. Brian Josephson theoretically studied the properties of supercurrent through a tunnel barrier and made a hypothetical prediction for superconductors that is commonly called Josephson effect (see details in the last portion of this chapter).

- Later on, in 1986, IBM researchers G. Bednorz and K. A. Muller succeeded to create a superconductor ceramic compound with an amazing critical temperature of 30K although ceramics are isolators as compared to other ceramic materials which was an interesting area for researchers at that time. Both the aforementioned researchers got the Nobel Prize in 1987 for their discovery of the first cuprate superconductor with a critical temperature of 35K. This discovery opened new horizons of research at high temperature leading to enormous number of publications during 1986 to 2001 [15].

- In 2003, Abrikosov was awarded the Nobel Prize for his revolutionary work on the theory of superconductors and super fluids. Several works have been reported on superconductors which are difficult to be summarised here [16,17].

- After a concise super past of Superconductors, it is necessary to address some relevant terms regarding superconductors in order to understand the nature of the super phenomenon of superconductivity.

3. Superconductors vs perfect conductors

In electronics, superconductors and perfect conductors are the two commonly used but quite confusing terms that is why they are usually misunderstood. In a real sense, the two terms are quite dissimilar [18].

- Superconductors exist in reality, or superconductors are approachable materials in reality while a perfect conductor is just a conceptual approach.

- Super conductor is an actual indication of zero resistivity as compared to perfect conductor that is an assumption and sometimes it is used to simplify the calculations and designs where the resistivity is quite insignificant.

- In order to be a superconductor, subcritical temperature is quite an important factor whereas a perfect conductor can have zero or low resistivity at any temperature.

- A lot of external factors like critical temperature, critical magnetic field etc. are necessary to achieve the super conductance but a perfect conductor does not need any external factor to achieve the perfect conductance or this is not the case for perfect conductor [19].

4. Phenomenon of superconductivity

If we talk about superconductivity, it is an extensively studied quantum mechanical phenomenon in solid state physics that is an outstanding combination of magnetic and electric properties below the critical temperatures. A superconductor experiences the Meissner Effect when kept in a magnetic field. This effect is related to removal of magnetic fields when a superconductor is cooled down to the point of its critical temperature. This is perhaps the most essential property of superconductors and pointed towards the zero resistivity [20].

In 1933, this Meissner effect was discovered by the physicists named W. Meissner and R. Ochsenfeld while working on lead as well as tin materials. So according to the Meissner effect, superconductors show zero (or close to zero) resistivity to electrical currents; moreover, they are perfect diamagnet [21]. Amazingly, some superconducting samples may attract magnetic fields in quite a few recent experiments that is the so-called paramagnetic Meissner effect [22].

Superconductors - Materials and Applications
Materials Research Foundations 132 (2022) 1- 16

Materials Research Forum LLC
https://doi.org/10.21741/9781644902110-1

In order to understand the above-mentioned phenomenon of superconductivity, it is important to address the following terms:

- Zero resistance
- Super-electron
- Critical temperature

4.1 Zero resistance

Zero resistance is the commonly used term for superconductors. Only those materials come under the heading of superconductors which fulfil this condition of zero resistance. Experiments have been done for the confirmation whether resistance is indeed zero or if there is any small residue of resistance. Gallop was the first scientist who finally justified the zero resistance of superconductors otherwise it is not possible to measure the current circulating in a superconducting loop by implanting an ammeter into the loop because current would definitely decay due to the resistance of the ammeter [23].

Moreover, the current in the loop and magnetic field are directly linked with each other so magnetic field quantification is possible without consuming energy from the circuit. This type of repetition of experiments over periods of years, confirmed the constant value of superconducting current and this persistent *current* is the main characteristic of superconductors which proves the zero resistance of superconductors.

For superconductors, it is quite preferable to define them in terms of R*esistivity* rather than conductivity since Resistivity is the reciprocal of conductivity (σ)

$$\rho = \sigma^{-1}.$$

So, mostly the superconductors are described by $\rho = 0$ rather than by $\sigma = \infty$.

As discussed under previous heading that it has been deducted from the lack of any decay of the current that the resistivity ρ of a superconductor metal is less than 10^{-26} Ω m whereas the resistivity of copper is 10^{-8} Ω m at room temperature. It is clear that the resistivity of superconductor metal is 18 times less than the resistivity of copper that is a normal conductor at room temperature [24]. Hence, it seems that we are justified in treating the resistivity of a superconducting metal as zero.

4.2 Super-electron

The electrons that cause the phenomenon of superconductivity are called super-electrons. It is assumed that their resistivity is nearly zero regarding that they are succeeding one another without any type of collision. They act in a vacuum. For resistance to less electrons, it is necessary that the current must be the same towards the path of electrons. The product of electron density and electron velocity may maintain the constant current. But in superconductors ions of metals are fixed, therefore, the electron density cannot be varied. In this way the current is the same and electrons don't have an electric field and cannot be accelerated [25].

4.3   Critical temperature for superconductors

The temperature at which the resistivity of superconductors vanishes is called critical or transition temperature of the superconductors [26]. Its value is not specific, i.e., varies from metal to metal and depends on the material's purity. Some metals are so sensitive that if little amount of impurity is present then they will not exhibit the phenomenon of superconductivity like iridium and molybdenum [27]. But this imagination is not for all metals like Cu and Na. This concept reveals the new type of superconductors. There are many metallic elements and alloys showing the property of superconductors. Some two metallic elements are not superconductors itself but their combination in the form of alloy exhibit the superconductivity phenomenon [28].

5.   Classification of superconductors

Those materials which exhibit superconductivity when sufficiently cooled are called superconductors. It has been considered for many years that the basic behavioural pattern of all superconductors is almost similar but in 1957 theoretical investigations of Abrosov led to the classification of superconductors that is based on the fact that superconductivity can vanish through two distinct situations. In other words, the actual basis of division of superconductors are the differences in their magnetic behaviour [29]. So, their classification as follows:

- Type-I superconductors,
- Type-II superconductors

Although the two classes have many similar properties even then the differences are enough for the bifurcation of superconductors into two distinct classes [30].

A brief tabulated difference of the aforementioned classes is as follows (Table 1):

Table 1. Difference between Type 1 and Type-II superconductors

Type-I superconductors	Type-II superconductor
These are called soft superconductors because low intensity magnetic fields can destroy their superconductivity.	These are called hard superconductors because it's not easy to destroy their superconductivity by an external magnetic field.
They are usually pure elemental superconductors like pure metals (except niobium, vanadium, technetium and carbon nanotubes). Some examples are Zn, Hg, Pb, Sn, Ta etc.	They are usually almost impure and compound superconductors like alloys and high critical temperature ceramics e.g., Nb_3Sn, Bi-Pd, Nb-Ti etc.

The conductivity of type-I superconductors is normally explained by BCS theory.	The conductivity of these type-II superconductors cannot be explained by BCS theory.
These are strongly diamagnetic in nature.	These are partially diamagnetic.
These are low temperature superconductors. (0-10K)	These are high temperature superconductors. (Above 10K)
These are used to prepare electromagnets. [31]	These cannot be used to prepare electromagnets.
This type of superconductors strongly follows the Meissner effect	These do not perfectly obey the Meissner effect but somewhat follow this effect.
They have low value of critical temperature and critical magnetic field (up to 1T)	They have a high value of critical temperature and magnetic field (greater than 1T).

6. Properties of superconductor

The fundamental properties of superconductors are as follows….

- Evanesce of the electrical resistance
- Diamagnetism as well as Flux lines
- Quantization of flux in superconductors
- Quantum interference
- Josephson currents

6.1 Evanesce of electrical resistance

The resistivity of a superconductor e.g., mercury suddenly decreases when it reaches a superconducting state. The basic method of resistance was used and voltage was measured when passing the electric current. Experimentally it is not possible to exactly prove the resistance value equal to zero, however, its upper range can be found.

For this purpose, highly sensitive methods were used to get the minimum possible residual resistance. In 1914 Kamerlingh-Onnes experimented with the decay of flow of electric current in a closed ring made of superconductor e.g., lead. The ring is kept in a normal state of temperature which is above the transition temperature and a magnetic rod adjusts in the ring-opening. Now down the temperature from the transition temperature T_c at which the ring becomes a superconductor by keeping magnetic field constant and sudden removal of magnetic rod induced electric current. Since the change in magnetic flux Φ caused, electric voltage then generates an electric current. Assumed that ring diameter is 5 cm with 1 mm of wire thickness, self-induction L coefficient is 1.3×10^{-7}H and the current decreasing by 1% in an hour then applying exponential decay law resulting there would be a change in magnitude of superconducting state more than eight orders [32].

Kamer-Lingh as well as Tuyn used another setup that contained 2 superconducting rings which produced a permanent flow of current. The ring present inside is held by a torsional strand and to some extent moved far-off its place, it seems that permanent current attracts thread to its side resulting in equilibrium in angular momentum of thread and permanent current. The equilibrium observed via light beam indicates no change in permanent current has been found [33].

To monitor the upper range of resistance in a superconducting state geometrically depends on self-induction coefficient L which value is required to be very small and time of observation. In the present time, it is known that the resistance jump during superconductor entry is at least 14 orders of magnitude. A superconductor can have a maximum of 17 orders of magnitudes of electrical-resistance that is lesser than the specific resistance of copper. From the above debate, finally it is inferred that electrical resistance vanishes in the superconducting state [34].

6.2 Flux lines and diamagnetism

In 1924 Kamerlingh-Onnes experimented with the magnetic behaviour of superconductors. He cooled down the lead-made hollow sphere to the transition temperature and applied the external magnetic field, then off the external magnetic field considering R equals zero. Keeping in view the history, applying the same conditions to the material which could be moved into the various states. It concludes that there would be exactly no single superconducting phase but different phases along arbitrary shielding currents [35]. In contrast to the ideal electrical conductors, the superconductor behaves differently [36].

Reassuming a sample cooled to the critical temperature and applied a very small magnetic field then the field escaped from the inner of sample excluding the outer coating of material resulting as an ideal diamagnetic state [37]. It was first discovered by Meissner and Ochsenfeld in 1933 and the phenomenon was observed on lead or Tin made rods. The experiment was done on a permanent magnet placed on a lead bowl having the magnetic field applied externally when $T > T_c$ and cool down to the critical temperature. On reaching the critical temperature $T < T_c$ then permanent magnet expelled by the magnetic field and raised from the lead bowl to the position gained after equilibrium. The image shown above is for reference. Because two types exist in superconductors, their behaviours relevant to the magnetic field are also different [38].

- Type I superconductors e.g., lead and mercury eject magnetic field to the critical field which converts superconductor to the normal conducting state when applied to the field of larger magnitude.

- Type II superconductors e.g., mostly alloys show ideal diamagnetism at lower critical magnetic field B_{c1} smaller than the usual magnetic field. To apply the upper critical magnetic field B_{c2} causing dissipation of superconducting property.

However critical fields in both type I and type II superconductors reach zero on reaching the critical temperature. Keeping in view the behaviour of superconductors of type-II at the lowest and the upper critical limits of magnetic-field subsequently also known as Shubnikov phase passes the shielding current and concentrates the magnetic field to generate a flux line system known as Abrikosov Vortices [39]. He got a Nobel prize in 2003 for studying this quantized phenomenon of flux-lines. Ideally, the flux line system is arranged in a superconductor in an equilateral triangular manner. The flux lines consisted of circulating current in combination with externally applied magnetic-field to form magnetic-flux resulting in a decrease in magnetic field between flux-lines. Conclude that increasing the external magnetic field in the superconductor causes the decrease in distance between flux lines [40].

6.3   Flux quantization in superconductors

A permanent current can be induced by the superconducting ring in the presence of the magnetic field. Magnetic current can be calculated via $\Phi=LI$. Concerning the macroscopic studies, the permanent current of any value can be induced with proper selection of magnetic field and any arbitrary value of magnetic flux from the ring can be taken. The magnetic-field of flux-lines carries flux quantum Φ_o. Here flux quantum is significant for a superconductor's performance. This idea was first taken by Fritz London in 1950 based on probability [41].

The experiment on the superconducting hollow cylinder was done to measure the magnetic flux quantization in 1961. It was worked and published by the two groups of physicists at Stanford; the first group including members was Munich, Deaver, and Fairbank and the second one's name was Doll and Nabauer, Experiment, the lead made a small hollow cylinder of 10μm evaporated to quartz rod was used and permanent current induced by cooling the small superconducting ring (known as freezing field B_f) by keeping it in parallel position to axis of cylinder. After reaching the critical temperature off the field resulting in a permanent magnet [42]. The calculation of flux in frozen may be calculated via torque applied on perpendicular position of field and cylinder axis which is why the sample is attached with the quartz rod. The light indicator and mirror were used for the deflection. The further self-resonance technique was used by the physicists because the torque value obtained in the case was too small. They excite the torsional oscillation of the system by using a photocell and a light beam so the field follows the resonance frequency which would reverse periodically at the frequency of oscillation resulting in the large amplitude of torque measured [43].

Deaver and Fairybank used a superconducting hollow cylinder for the determination of elementary flux quantum Φ_o. They find the frozen-in-flux differently. They used a small detector coil and the oscillation end to end along the axis with a frequency at 100 Hz resulting in the generation of inductive voltage. The voltages were then further amplified to consider their value for measurement [44].

6.4   Quantum interference

The effect of coherent matter-wave in a superconductor is another aspect of study and demonstrated by diffraction and interference. Interference is the phenomenon in which light passes through a double slit and projects to the screen. Sagnac Interferometer, a laser beam from the source ejected, passed through the semi-transparent mirror and splits into two beams travelling in the spherical path by opposite direction, three furthermore mirrors adjusted circularly [45].

If both waves with the same phase reached the detector, then a large signal would be obtained due to constructive interference. If the setup moves in a clockwise direction, then the mirror rotates in the opposite direction of the beam. As a result, the ray has to cover more distance than before to reach the sensor compared to the ray that travels in a counter clockwise direction. Due to this behaviour, the phase difference aroused in the detector causes aggression in the rotational velocity of measurement setup which then affects the periodic signal gain between upper and lower value [46].

To overcome this issue, the Sagnac interferometer as a gyroscope can be used. The phenomenon can occur in superconductors using coherent matter waves which can be further elaborated via the Josephson effect.

6.5   Josephson current

Two superconductors are placed up and down in the form of a sandwich and there is minor space between them also called a non-superconductor insulator or barrier. When the insulator is too small, electrons from one superconductor can move to the other superconductor due to quantum mechanical tunnelling. Wavefunction works here briefing the reason for moving or leaking electrons from the metal area. when another metal is passed near the barrier, then the current can flow between the sandwich structure [47, 48]. Due to the flow of electrons between superconductors, a weak supercurrent passes through the barrier. This weak supercurrent is also known as *Josephson current* which was predicted by Brain D. Josephson in 1962 and got a Nobel prize in 1973 due to the discovery of amazing properties related to superconductors. Properties of the Josephson junction are very important for the elaboration of macroscopic-wave-function that is proportional to Josephson current [49]. As discussed above sandwich form of superconductors, consider the same scenario with leakage of coherent matter-wave from superconductors and for this purpose, the Josephson equation has been derived as follow:

$$\frac{d}{dt}(\varphi_2 - \varphi_1) = \frac{2eU}{\hbar} = \frac{2\pi}{\Phi_0} U$$

Most barriers and superconductors follow the Josephson equations and these equations are valid in most cases. The junctions, elaborate as SIS junction is superconductor-insulator-superconductor in which insulator should be 1-2nm thick SNS junction is superconductor-normal conductor-superconductor, it operates with high thickness of normal conductor due to cooper pair which have less penetration in oxide layer than normal conductors.

Materials Research Foundations 132 (2022) 1- 16

https://doi.org/10.21741/9781644902110-1

Moreover, resistance is the important factor between normal conductors and oxide junctions; normal conductors have less resistance (10^{-8} per square) than oxide junctions ($10^{-4} - 10^{-3}$ per square) [50, 51].

Conclusion

Finally, it may conclude that superconductors are the materials which behave as normal conductors at standard values of room temperature and pressure. They show resistance in the flow of current but on cooling below their critical values of temperature (above the absolute zero) and magnetic field, they exhibit the rare phenomenon of super conductance i.e., zero resistivity. These specific values of critical temperature and critical magnetic field at which conductors become superconductors principally depend upon the nature of super conducting material. However, there is a difference between perfect conductor and super conductor where the former one is ideal and does not exist in real, but the later one is practically approachable. Since the discovery of superconductors by Onnes in 1911, a lot of developments have been done in this field to date and few scientists got noble prize for their extra ordinary work on superconductors. The main properties of superconductors are the disappearance of electrical resistance and perfect diamagnetism. Flux quantum as well as Josephson equations are also significant for superconductors.

References

[1] J. Bardeen, L. N. Cooper, J. R. Schrieffer, Theory of superconductivity, Phys. Rev. 108 (1957) 1175. https://doi.org/10.1103/PhysRev.108.1175

[2] J. Bardeen, Electron-vibration interactions and superconductivity, Rev. Mod. Phys. 23 (1951) 261. https://doi.org/10.1103/RevModPhys.23.261

[3] P. G. De Gennes, P. A. Pincus, Superconductivity of metals and alloys, CRC Press, (2018). https://doi.org/10.1201/9780429497032

[4] J. Hara, K. Nagai, Superconducting transition-temperature of thin-films in magnetic-field, J. Phys. Soc. Jpn. 63 (1994) 2331-2336. https://doi.org/10.1143/JPSJ.63.2331

[5] M. R. Beasley, A History of Superconductivity. In Advances in Superconductivity, Springer, Tokyo, 1989. https://doi.org/10.1007/978-4-431-68084-0_1

[6] J.G. Bednorz, K. A. Muller, Earlier and recent aspects of superconductivity, 1990. https://doi.org/10.1007/978-3-642-84377-8

[7] E.H. Brandt, The flux-line lattice in superconductors. Rep. Prog. Phys. 58 (1995) 1465. https://doi.org/10.1088/0034-4885/58/11/003

[8] P. Jung, A. V. Ustinov, S. M. Anlage, Progress in superconducting metamaterials, Supercond. Sci. Technol. 27 (2014) 073001. https://doi.org/10.1088/0953-2048/27/7/073001

[9] E. M. Towsif, Analysis of Prospective Elements and Crystal Lattice Structures via Computer Algorithms to Identify Standard Temperature Pressure (STP) Superconductors, arXiv preprint arXiv. (2021) 2110.15201.

[10] P.F. Dahl, K. Onnes and the discovery of superconductivity: The leyden years, 1911-1914, Hist. Stud. Phys. Sci. 15 (1984) 1-37. https://doi.org/10.2307/27757541

[11] R. L. Fagaly, Superconducting quantum interference device instruments and applications, Rev. Sci. Instrum. 77 (2006) 101101. https://doi.org/10.1063/1.2354545

[12] V. Kozhevnikov, Meissner Effect: History of Development and Novel Aspects. J. Supercond. Nov. Magn. 34 (2021) 1979-2009. https://doi.org/10.1007/s10948-021-05925-8

[13] W. Meissner, R. Ochsenfeld, Ein neuer effekt bei eintritt der supraleitfähigkeit, Naturwissenschaften, 21 (1933) 787-788. for English translation see A. M. Forrest, Meissner and Ochsenfeld revisited, A new effect concerning the onset of superconductivity, Eur. J. Phys. 4 (1983) 117. https://doi.org/10.1007/BF01504252

[14] P. J. Ford, G. A. Saunders, The rise of the superconductors, CRC press, 2004. https://doi.org/10.1201/9780203646311

[15] Goodstein, David, J. Goodstein, Richard Feynman and the history of superconductivity, Phys. Perspect. 2 (2000) 30-47. https://doi.org/10.1007/s000160050035

[16] C. M. Rey, A. P. Malozemoff, Fundamentals of superconductivity, In Superconductors in the Power Grid, Woodhead Publishing, 2015, pp. 29-73. https://doi.org/10.1016/B978-1-78242-029-3.00002-9

[17] J. Orenstein, A. J. Millis, Advances in the physics of high-temperature superconductivity, Sci. 288 (2000) 468-474. https://doi.org/10.1126/science.288.5465.468

[18] V. V. Schmidt, V. V. Schmidt, P. Müller, A. V. Ustinov, The physics of superconductors: Introduction to fundamentals and applications. Springer Science & Business Media, 1997. https://doi.org/10.1007/978-3-662-03501-6_1

[19] F. S. Henyey, Distinction between a Perfect Conductor and a Superconductor. Phys. Rev. Lett. 49 (1982) 416. https://doi.org/10.1103/PhysRevLett.49.416

[20] J. R. Schrieffer, Theory of superconductivity, CRC press, 2018. https://doi.org/10.1201/9780429495700

[21] L. Greene, T. O. M. Lubensky, M. Tirrell, P. Chaikin, H. Ding, K. Faber, S. Zinkle, Front. Mater. Research, A Decadal Survey (No. Doe-Nasem-16257). The National Academies of Sciences, Engineering and Medicine, 2019. https://doi.org/10.2172/1556101

[22] F. Parhizgar, A. M. Black-Schaffer, Diamagnetic and paramagnetic Meissner effect from odd-frequency pairing in multiorbital superconductors, Phys. Rev. B. 104 (2021) 054507. https://doi.org/10.1103/PhysRevB.104.054507

[23] K. Sakamaki, H. Wada, H. Nozaki, Y. Ōnuki, M. Kawai, Carbosulfide superconductor. Solid State Commun. 112 (1999) 323-327. https://doi.org/10.1016/S0038-1098(99)00359-2

[24] Huse, A. David, M. Fisher, D. S. Fisher, Are superconductors really superconducting? Nature 358 (1992) 553-559. https://doi.org/10.1038/358553a0

[25] L. R. Tagirov, Proximity effect and superconducting transition temperature in superconductor/ferromagnet sandwiches, Physica C. Superconductivity, 307 (1998) 145-163. https://doi.org/10.1016/S0921-4534(98)00389-X

[26] G. Burns, High-temperature superconductivity, Elsevier Science & Technology, 1992.

[27] G. Krabbes, G. Fuchs, W. R. Canders, H. May, R. Palka, High temperature superconductor bulk materials. Fundamentals-processing-properties control-application aspects. 2006. https://doi.org/10.1002/3527608044

[28] V. Z. Kresin, S. A. Wolf, Fundamentals of superconductivity: Springer Science & Business Media, 2013.

[29] M. Strongin, A. Paskin, D. G. Schweitzer, O. F. Kammerer, P. P. Craig, Surface superconductivity in type I and type II superconductors. Phys. Rev. Lett. 12 (1964) 442. https://doi.org/10.1103/PhysRevLett.12.442

[30] T. Yogi, G. J. Dick, J. E. Mercereau, Critical rf magnetic fields for some type-I and type-II superconductors, Phys. Rev. Lett. 39 (1977) 826. https://doi.org/10.1103/PhysRevLett.39.826

[31] W. Buckel, R. Kleiner, Superconductivity: fundamentals and applications, John Wiley & Sons, 2008.

[32] D. J. Quinn III, W. B. Ittner III, Resistance in a Superconductor, J. Appl. Phys. 33 (1962) 748-749. 32 https://doi.org/10.1063/1.1702504

[33] P. F. Dahl, Kamerlingh onnes and the discovery of superconductivity, The leyden years, 1911-1914, Hist. Stud. Phy. Sci. 15 (1984) 1-37. https://doi.org/10.1016/0039-3681(84)90027-X

[34] F. Lacy, Using Electromagnetic Properties to Identify and Design Superconducting Materials, 2021. https://doi.org/10.5772/intechopen.97327

[35] L. P. Lévy, Magnetism and superconductivity: Springer Science & Business Media, 2000.

[36] D. Shoenberg, Magnetic Properties of Superconductors, Nature, 142 (1938) 874-875. https://doi.org/10.1038/142874d0

[37] A. Bussmann-Holder, H. Keller, High-temperature superconductors: underlying physics and applications, Z. Naturforschung B, 75 (2020) 3-14. https://doi.org/10.1515/znb-2019-0103

[38] V. Cvetkovic, Z. Tesanovic, Multiband magnetism and superconductivity in Fe-based compounds, Europhys. Lett. 85 (2009) 37002. https://doi.org/10.1209/0295-5075/85/37002

[39] Brandt, E. Helmut, The flux-line lattice in superconductors, Rep. Prog. Phys. 58 (11) (1995) 1465. https://doi.org/10.1088/0034-4885/58/11/003

Superconductors - Materials and Applications   Materials Research Forum LLC
Materials Research Foundations 132 (2022) 1- 16   https://doi.org/10.21741/9781644902110-1

[40] P. Nozières, W. F. Vinen, The motion of flux lines in type II superconductors, The Philosophical Magazine, J. Theor. Appl. Phys. 14 (1966) 667-688. https://doi.org/10.1080/14786436608211964

[41] H. A. Mook, M. D. Lumsden, A. D. Christianson, S. E. Nagler, B. C. Sales, R. Jin, C. D. Cruz, Unusual relationship between magnetism and superconductivity in Fe. Te. 0.5 Se 0.5, Phys. Rev. Lett. 104 (2010) 187002. https://doi.org/10.1103/PhysRevLett.104.187002

[42] C. E. Gough, M. S. Colclough, E. M. Forgan, R. G. Jordan, M. Keene, C. M. Muirhead, S. Sutton, Flux quantization in a high-Tc superconductor, Nature, 326 (1987) 855-855. https://doi.org/10.1038/326855a0

[43] V. L. Ginzburg, Magnetic flux quantization in a superconducting cylinder, Sov. Phys. JETP, 15, (1962) 207-209.

[44] R. Doll, M. Näbauer, Experimental proof of magnetic flux quantization in a superconducting ring, Phys. Rev. Lett. 7 (1961) 51. https://doi.org/10.1103/PhysRevLett.7.51

[45] R. L. Fagaly, Superconducting quantum interference device instruments and applications, Rev. Sci. Instrum. 77 (2006) 101101. https://doi.org/10.1063/1.2354545

[46] R. Kleiner, D. Koelle, F. Ludwig, J. Clarke, Superconducting quantum interference devices, State of the art and applications, Proceedings of the IEEE, 92 (2004) 1534-1548. https://doi.org/10.1109/JPROC.2004.833655

[47] Tanaka, Yukio, S. Kashiwaya, Theory of Josephson effects in anisotropic superconductors, Phys. Rev. B, 56 (1997) 892. https://doi.org/10.1103/PhysRevB.56.892

[48] M. V. Feigel'man, L. B. Ioffe, Microwave properties of superconductors close to the superconductor-insulator transition, Phys. Rev. Lett. 120 (2018) 037004. https://doi.org/10.1103/PhysRevLett.120.037004

[49] P. Seidel, Josephson effects in iron-based superconductors, Supercond. Sci. Technol. 24 (2011) 043001. https://doi.org/10.1088/0953-2048/24/4/043001

[50] T. Matsushita, Flux pinning in superconductors (Vol. 164). Berlin: Springer, 2007.

[51] T. Matsushita, Flux Pinning Phenomena: In Superconductivity and Electromagnetism, Springer, Cham., 2021, pp. 69-113. https://doi.org/10.1007/978-3-030-67568-4_5

Superconductors - Materials and Applications
Materials Research Foundations 132 (2022) 17-48

Materials Research Forum LLC
https://doi.org/10.21741/9781644902110-2

Chapter 2

Properties and Types of Superconductors

M.S. Hasan[1] and S.S. Ali[2]*

[1]Department of Physics, The University of Lahore, Lahore 54000, Pakistan

[2]School of Physical Sciences, University of the Punjab, Lahore 54590, Pakistan

*shahbaz.sps@pu.edu.pk

Abstract

The disappearing of electrical resistance below the critical temperature (T_c) is known as superconductivity discovered by Kamerlingh Onnes in 1911. Superconductors are consisted of two categories namely type I and type II also called soft and hard superconductors, respectively. Type I superconductors obey the Meissner effect while type II superconductors do not. The superconducting compounds are divided into three categories, (i) Metal based systems, (ii) Copper oxides (cuprates) and (iii) Iron based superconductors (IBSC). Metal-based superconductors are combination of cubic crystal configuration named the A15 structure. Initial IBSC was revealed in 2006 for LaFePO; nevertheless, T_c stayed as small as ~4 K. High-T_c compounds were then expressed for $LaFeAsO_{1-x}F_x$ by means of $T_c = 26$ K in 2008. Superconductors have considerable positions in the lower temperature magnet applications such as MRI, nuclear magnetic resonance and superconducting quantum interference.

Keywords

Superconductors, Types of Superconductors, Meissner Effect, Metal Based Systems, Cuprates, IBSC

Abbreviations used

T_c	Critical/transition temperature
AFM	Anti ferromagnetic
IBSCs	Iron based superconductors
H	Magnetic field
BCS	Bardeen, Cooper and Schrieffer
H_c	Critical magnetic field
H_{c2}	Upper critical field
H_{c1}	Lower critical field
H_{c3}	Higher field
PIT	Power in tube

BSCCO	BiSrCaCuO
YBCO	Y–Ba–Cu–O
LED	Light emitting diode
β	Electron band
α	Centered zone
Γ	Point of the Brillouin zone
HTSC	High-temperature superconductors
ΔT	Super cooling temperature
c-GSs	c-axis growth sectors
SFCL	Superconducting fault current limiters
U	Effective energy

Contents

1.   Introduction

The entropy decreases with the decrease in temperature and materials obtain ordering state due to thermodynamic equilibrium. At absolute zero temperature the materials become completely ordered. The order state continues during phase transitions. The significant phase transitions (few Kelvin temperatures) lead these materials to superconducting phases. Frictionless flow is demonstrated by these phases at superficial level [1].

The stage conversion engages the electron fluid which is liable for electrical conductivity normally at high temperature for superconducting metals. The frictionless flow is possible due to very small electrical resistivity (unmeasurable). The scattering of conduction electrons causes the electrical resistivity because of defects in crystal lattice. So it can be considered that scattering process in superconductors is strangely turned off [2].

The Meissner effect is revealed by all superconductors, according to this effect in presence of small magnetic field they act as diamagnetic materials. If the magnetic flux is absent during reversible manner and then they perform just similar to conservative diamagnetic materials. Rather they have huge diamagnetic susceptibility [3].

1.1   The Meissner effect and superconductors

Meissner effect was discovered in 1933 and was initial evidence for paradoxical effect [4]. The thermodynamic cannot explain the individual atoms as second law of thermodynamics cannot be challenged by quantization of electron orbits. The rings used during works are bodies composed of huge amount of atoms, electrons or cooper couples [5-8]. Hence law of entropy enhance is confronted by quantization in rings [9]. Discreteness of the bands of superconducting ring is associated to the numeral of cooper couples, mathematically [10],

$$N_s = 2\pi r s n_s \qquad (1)$$

Quantization is examined in macroscopic superconducting configuration, and superconductivity is macroscopic quantum fact. Heike Kamerlingh was the first man who discovered superconductivity in 1911 by observing the decrease in resistance to zero at low temperature. Before 1933 the scientists believed that electric current only emerges in accord with thermodynamics laws and Newton second law as,

$$m\frac{dv}{dt} = qE \qquad (2)$$

According to these laws the current density in ideal conductor

$$j = n_s qv \qquad (3)$$

Where, n_s is density of mobile carriers of charge q. It is altered by the function of electric field E in time as,

$$\frac{dj}{dt} = \left(\frac{n_s q^2}{m}\right) E \tag{4}$$

$$= \frac{E}{\mu_0 \lambda_L^2} \tag{5}$$

The Maxwell's relations should be valid as,

$$rot\, E = -\frac{dB}{dt} \tag{6}$$

$$rot\, H = j \tag{7}$$

$$B = \mu_0 H \tag{8}$$

$$\lambda_L^2 \nabla^2 \frac{dH}{dt} = \frac{dH}{dt} \tag{9}$$

According to the above relation, the transformation in magnetic field (H) with respect to time is able to enter into ideal conductor with the penetration depth λ_L. In extensive cylinder with macroscopic radius (R >> λ_L) then magnetic field inside it,

$$h = He^{-\frac{R-r}{\lambda_L}} \tag{10}$$

By escalating the exterior magnetic field from zero to H at T < T_c then the density of surface screening current will be as,

$$j = j_0 e^{-\frac{R-r}{\lambda_L}} \tag{11}$$

Where, j_0 is current density (H/λ_L) at r = R. Kinetic energy of screening current as,

$$E_k = \mu_0 H^2 \lambda_L \pi RL \tag{12}$$

$$= \mu_0 H^2 V \frac{\lambda_L}{R} \tag{13}$$

The density corresponding to energy of macroscopic cylinder having radius R >> λ_L and length L is as,

$$\varepsilon = \frac{n_s m V^2}{2} \tag{14}$$

Superconductors - Materials and Applications
Materials Research Forum LLC
Materials Research Foundations 132 (2022) 17-48
https://doi.org/10.21741/9781644902110-2

$$= \frac{\mu_0 \lambda_L^2 j^2}{2} \tag{15}$$

The scientists considered that after the evolution in normal metal position having resistivity $\rho > 0$, the surface screening current have to be dissipated in joule heat contained by the limited interval. Also, the current will disappear as the cylinder goes back to superconducting position at $\rho = 0$ in applied magnetic field H during interval dH/dt = 0 [11].

According to W. Meissner and R. Ochsenfeld, in 1933 they found that exterior screening current emerges during shift into superconducting position in presence of H during interval dH/dt = 0, opposing the Newton second law and electrodynamics laws [4]. Experimentally, the superconductors are different from ideal conductors, opposes to law of entropy enhance if the exterior currents related by means of field are damping with the production of Joule heat [11] while superconductivity is demolished. According to the scientists, the Meissner effect is proficient to twist the irreversible conversion in which Joule heat is produced into a reversible transition in which no Joule heat is. It remains mystery that the determined currents are exterminated without production of Joule heat. However, in 1933 the scientists reached a point that superconducting transition is phase transition. They were convinced that it is reversible thermodynamic procedure without Joule heating. All theories related to superconductivity were designed on basis of this effect.

2. History of superconductors

Disappearing of electrical resistance under critical temperature (T_c) is known as superconductivity. Kamerlingh Onnes in 1911 discovered the superconductivity. For many decades the superconductivity was severe challenge for the theoretical physics. In 1935, foremost successful combination of phenomenological mathematical relations for superconducting materials was demonstrated by F. London. Till to 1950, following the detection of this fact, there was no sufficient microscopic theory of superconductivity. Conversely, particular elements essential for successful theory to describe superconductivity was known to theorists in 1935. In 1925, unusual reduction of a Bose-Einstein gas was explained by Einstein. Since 1931, concept about couples of fermions could be merged to shape bosons was introduced. F. London gave famous concept of superconductivity in 1951in his book "Superfluids", volume 1. Finally in 1957, the initial theory to demonstrate the microscopic behaviour of superconductivity in metal and alloy named as BCS theory was given by (Bardeen et al., 1957).

Convenient functions of superconductivity are progressively getting better each day. On the other hand, definite exploit of superconducting gadgets is incomplete due to the reason that they have to be low temperatures to turn into superconducting. For instance, superconducting magnets in accelerators are cooled down by application of liquid helium, and, it is compulsory to utilize cryostats that have ability to produce temperatures of 4 K. As the helium resources are not vast, and supply of helium will vanish in the near future. Consequently, liquid nitrogen can be employed for high-T_c superconductors cooling as of

Superconductors - Materials and Applications   Materials Research Forum LLC
Materials Research Foundations 132 (2022) 17-48   https://doi.org/10.21741/9781644902110-2

abundance of nitrogen in the world. So, the superconductors having critical temperatures higher than the 77 K can be cooled down by using liquid nitrogen.

The microscopic behaviour of superconductivity in metals can be explained by the BCS theory (Bardeen et al., 1957). According to this theory, electrons in a metallic superconductor are coupled by the exchange of phonons. Various scientists like (De Jongh, 1988; Emin, 1991; Hirsch, 1991; Ranninger, 1994) claims that the BCS theory is not suitable to describe the behaviours of superconductivity in oxide superconductors. However, former representations depending upon BCS-like representation return the phonons by other bosons, for instance: plasmons, excitons and magnons, as the mediators rooted the smart interface amongst couple of electrons and various researchers stated superconductivity within oxide can be explained through the BCS theory or BCS-like theories (Canright & Vignale, 1989; Tachiki & Takahashi, 1988; Takada, 1993).

Significant high-T_c superconductors lie in the category of copper oxide superconductors. Innovation of room temperature superconductor must activate technical uprising. Although, the microscopic behaviour for higher T_c superconductors is indistinct till to present era. The purpose of this chapter is to demonstrate the fundamental and properties of superconductors.

3.   Types of superconductors

The superconductors are classified into two categories named as:

- Type I superconductors
- Type II superconductors

3.1   Type I superconductors

These superconductors are consisted of basic conductive divisions and are applicable in various parts like electrical cabling to microchips in computers and in coils for superconducting magnets. In an applied magnetic field these superconductors lose their superconductivity at critical magnetic field (H_c). After this it behaves as a conductor. Due to such property these are also called soft superconductors. Such superconductors obey the Meissner effect entirely. These superconductors demonstrate the magnetization of the form as given in Figure 1.

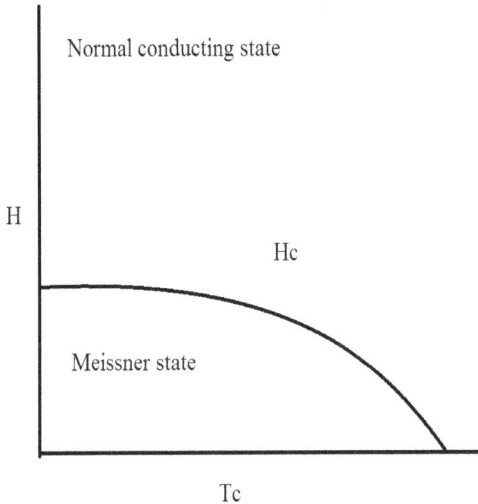

Figure 1: Magnetization curve for Type I superconductors, Meissner state is separated

3.1.1 Examples

Examples of type I superconductors are zinc and aluminium.

3.2 Type II superconductors

Such types of superconductors drop their superconductivity slowly after an external magnetic field is applied. At less considerable magnetic field such superconductors will drop their superconductivity and at higher critical magnetic field they will entirely drop their superconductivity. The situation amongst minor significant magnetic field and upper critical magnetic field is named as middle positioner vortex state. As these conductors lost their superconductivity slowly but not easily hence these are also called hard superconductors. Such superconductors do not obey the Meissner effect entirely. Such superconductors are potential candidate for powerful field superconducting magnets. Exhibited magnetization curve for type II superconductors is shown in Figure 2. Type II superconductors are leaned to be transition metals with great standards of electrical resistivity in usual situation. These have superconducting electrical characteristics capable of field signified by H_{C2}. Meissner effect is imperfect for the flux density B = 0, amongst the superior critical field (H_{C2}) and inferior critical field (H_{C1}).

Superconductors - Materials and Applications
Materials Research Foundations 132 (2022) 17-48

Materials Research Forum LLC
https://doi.org/10.21741/9781644902110-2

Normal conducting state

H_{C2}

H

Mixed state

H_{C1}

Meissner state

Tc

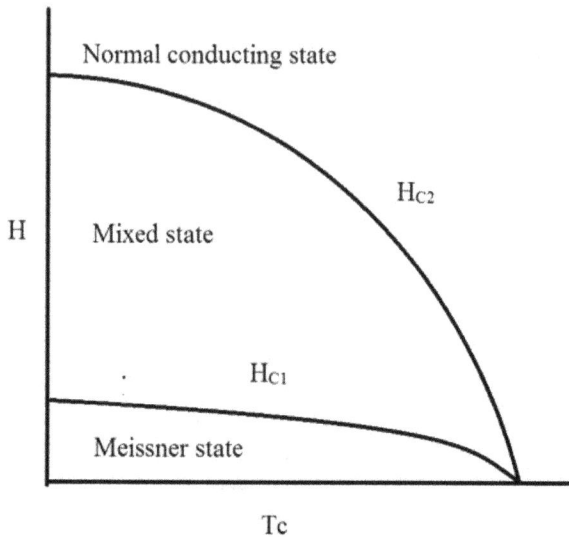

Figure 2: Magnetization curve for Type II superconductors. Meissner state, mixed state and normal conducting stateare separated by H_{C1} and H_{C2} lines by H_c

3.2.1 Examples

NbN and BaBi$_3$ are best examples for type II superconductors.

4. Comparisons between type I and type II superconductors

Both types of superconductors have similar mechanism of superconductivity. Also, both types of superconductors have identical thermal characteristics at normal transition in zero magnetic fields. Apart from the mechanism there are also other differences existing in both type of superconductors. The following are the differences between both types of superconductors.

4.1 Meissner effect

Meissner effect is the magnetic property for superconductors; this effect is dissimilar for both types of superconductors. Type I superconductors leave out the magnetic field till suddenly superconductivity is vanished, then the field enters entirely. Type II superconductors left out the magnetic field entirely till the lower critical field (H_{C1}). Beyond the lower critical field (H_{C1}) when the field is partially excluded, the superconductors maintain their superconducting properties. At upper critical field (H_{C2}) the superconductivity disappears as the flux penetrates fully. However, the outer most surface layer of superconductors may keep the superconducting at higher field (H_{C3}).

4.2 Conduction of electrons

The significant distinctions between type I and type II superconductors are signified open pathways of conduction electrons in the common position. If coherence length (ξ) is greater to the penetration depth (λ), then the superconductors will fall into the type I superconductor category. Mostly the pure metals belong to the type I, where,

$$K = \frac{\lambda}{\xi} < 1 \tag{16}$$

While the mean free path and coherence are shorter and penetration length is greater, then the superconductors will be type II. This situation occurs when,

$$K = \frac{\lambda}{\xi} > 1 \tag{17}$$

4.3 Surface energy

Suppose the interface amongst the superconducting state and normal state regions. Such interface contains surface energy either positive or negative and tends to decrease by means of the raise of a magnetic field. With the increase in the magnetic field the surface energy will be positive and superconductors will be type I. If the magnetic field is amplified and the surface energy is negative then superconductors will be type II. For transition temperature the mark of surface energy would have no significance.

5. Superconducting materials

In the present era the scientists have developed various materials that have ability to maintain their superconductivity at room temperatures [12]. The widespread explorations for new superconductors have been carried out to investigate the finest materials with high critical temperature (T_C). Even after passing hundred years of this discovery, the superconductivity is still a mystery, for this reason the scientists are searching innovative superconducting materials [13]. Superconducting materials are divided into three categories which are [14];

- Metal based system
- Copper oxides (cuprates)
- Iron based superconductors

The widespread investigation for new superconductors, in novel materials by means of greater critical temperature (T_c), are executed ever since [15]. Figure 3 [14] demonstrates the various superconducting materials for T_c values at ambient pressure during a number of years [14]. All superconducting materials are explained with the chronological manner [16]. Various types of superconducting materials along with critical temperatures are illustrated in Table 1.

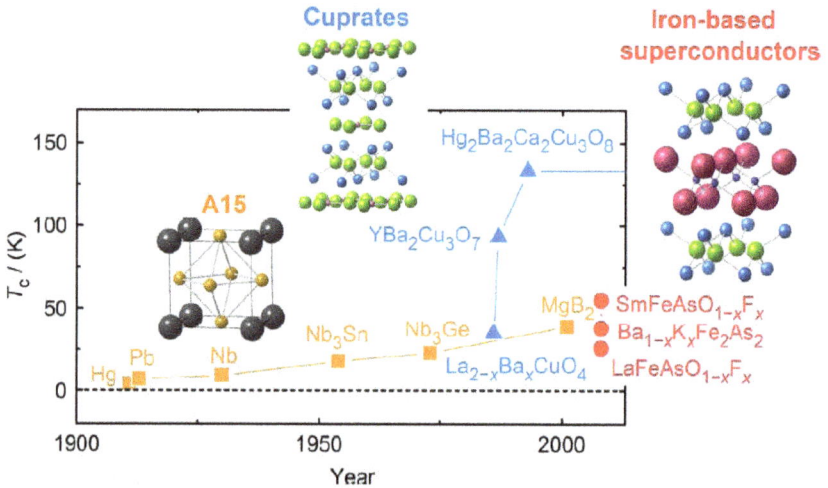

Figure 3. [14] The progress and crystal structures of superconductors

Table 1: Different superconducting materials with critical temperatures.

Materials	T_c (K)
Al	1.2
In	3.4
Sn	3.7
Pb	7.2
Nb	9.2
NbN	16
Nb_3Ge	23
$YBa_2Cu_3O_7$	92

5.1   Metal based system superconductors

The metal originated superconductors consist of cubic crystal arrangements named as A15. The Structure is given in Figure 4. In case of V3Si and Nb₃Ge the T_c was gradually enhanced via modification of the inserted elements in crystal arrangements. Berndt Matthias demonstrated 6 laws for discovery of novel superconductors [17]: (1) better the high symmetry while best is the cubic symmetry, (2) the electronic states in case of high density are better, (3) in the absence of oxygen, (4) avoided from the Mg, (5) insulators must not include, and (6) fail to attend from theorists. For exploration of metal originated superconductors the laws represented by Berndt are verification with the increase of T_c in the alloys of A15. Central procedure in A15 materials classification is a phonon-mediated method, due to the pairing of conduction electrons (named as Cooper coupling) with lattice vibrations. The controlled superconductors arranged through the procedure of phonon are called conventional superconductors. It is due to the fact that it is based on the BCS theory which was proposed by Bardeen, Cooper and Schrieffer in 1957. Earlier, they had given the explanation of microscopic behaviour for superconductivity. For the Nb3Ge superconductors the utmost T_c was illustrated approximately 23 K temperature. Various scientists considered that T_c was the intrinsic boundary organized by the BCS theory and gave the name BCS wall. Johannes Georg Bednorz and Alex Muller in 1986 discovered the novel superconductors lying on electron and phonon pairing beyond the T_c nearly 31 K in materials consisted of BaLaCuO coordination for the deposited perovskite formation [18]. The publication of article with the title "Possible High T_c Superconductivity in the Ba–La–Cu–O system" followed by existence of superconductivity in bulk were confirmed [19]. Such outcomes open the new doors for the global researchers for superconductors. The given passion taken is related to the reason that the value of T_c is greater as compared to the BCS wall was invented from the materials that do not obey the Matthias laws. Such materials belonged to the family of oxide without cubic arrangement of crystals and materials was derived from Mott insulator before doping with antiferromagnet (AFM) having Cu^{2+} $(3d^9)$ configurations [20].

The field of molecular metals and superconductors has expanded quickly since the first molecular metal was invented in the 20th century [21]. Recently, Snider *at el.,* have discussed carbonaceous superconducting materials at room temperature and at eminent pressure [22]. Although, they studied superconducting phenomena at very high pressure but it has opened new doors for the researchers to discover new interesting superconducting materials with distinguished properties. ThCr2Si2 type materials which are also named as 122-type class having 700 members, various physical and chemical characteristics are demonstrated by 122-type structured composites [12]. The ThCr2Si2 type materials have tetragonal crystal structure with 14/mmm, no. 139 space group, AT2X2 where, A is lanthanide or alkaline earth metal, T is transition metal and X is P, Se, Si, Ge or As class of composites are attractive due to important physical properties and superconducting features at higher transition temperatures [23, 24] and lower transition temperatures [25, 26], formation of superconductivity via doping [27-29], pressure induce superconductivity [30-32], magnetic [33] and antiferromagnetic properties [34]. Ban and Sikirica were the pioneers who realized about $ThCr_2Si_2$ type materials in 1965 [35], while APd_2P_2 type structures were first initiated in 1983 [36]. Currently, Blawat et al., experimentally designed $CaPd_2P_2$ and $SrPd_2P_2$ type superconductivity compounds at 1.0 K and 0.7 K transition temperatures, correspondingly [37]. They found that anti-bonding forces in Pd-P govern nearly at Fermi level which is important feature for increase in superconductivity. The important experimental findings of APd_2P_2 materials are higher expense of Pd and barriers associated with volatile character of P [14].

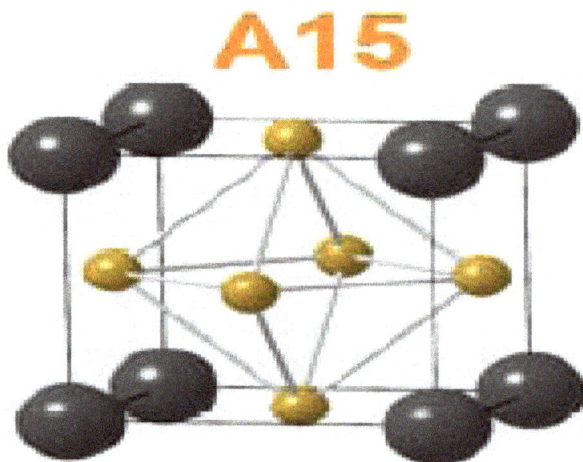

Figure 4 [14]: Structure of metal based superconductors

5.2   Copper oxides (Cuprates)

Two general properties are possessed by the high T_c crystal configuration of cuperate. Arrangement is consisted of CuO_2 planes named as structural unit. While the appearance of superconductivity is possible when charge carriers are doped in the form of insulating parent phase with antiferromagnetism. Even though, different novel superconductors are designed by the researcher and are reported lying in this category of cuprates without reporting of T_c since 1993 [38]. The combine configuration of cooper oxide ions are demonstrated in figure 5 [14]. In superconducting orthorhombic arrangements for Y and Bi classifications are existed. Furthermore, greater T_c cuprates expresses huge anisotropy for the characteristics of superconducting. 3D controlled crystals are required for the designing of superconducting wires and tapes. MgB_2 having 39 K T_c is high temperature intermetallic superconductor and was illustrated by Jun Akimitsu in 2001 [39]. For such

Materials Research Forum LLC
https://doi.org/10.21741/9781644902110-2

superconductors the investigations showed that the isotropic consequence and electronic heat capacity follow the conventional BCS procedure [40]. A detection of higher T_c within MgB_2 having uncomplicated structures and compositional formulas was amazing as the substance is primary metal originated superconductor apart from the 'BCS wall' [41]. Superconductivity comes into view exclusive of carriers doping, and predictable power-in-tube (PIT) process is employed to construct wires analogous with procedure applied for the fabrication of various metal alloy superconductors. The scientific shortcoming of the scheme is unavailability of significantly high magnetic field. Due to the lower doping other than the B isotope the utmost T_c will remain stable [14].

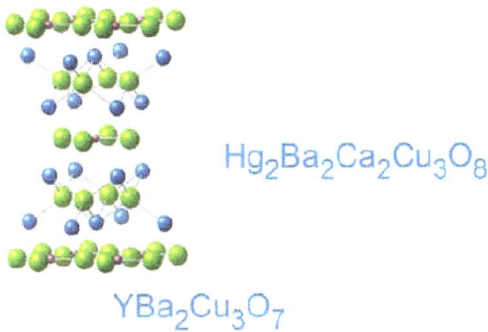

$$Hg_2Ba_2Ca_2Cu_3O_8$$

$$YBa_2Cu_3O_7$$

Figure 5 [14]: Structure of copper oxides (cuprates)

Cooper oxides or cuprates are second category of superconducting materials [42]. The cuprates demonstrates commendable features like stronger magnetic field presentation, nil energy failure and current bearing ability [43]. BSCCO (BiSrCaCuO) system is one of the wonderful compound that have get the intentions of scientists because of astonishing characteristics to carry the electric current with zero resistance. The cuprate system is consisted of three phases with common formula $Bi_2Sr_2Ca_nCu_nO_{2n+4+y}$ where, n = 1, 2, 3 taking into consideration the number of CuO_2 layer in sub lattice cell, correspondingly. Due to higher critical temperature ($T_c = 110$ K) the phases of Bi – 2223 are more significant than the Bi – 2201 ($T_c = 20$ K) and Bi – 2212 ($T_c = 90$ K) [44]. Various researchers made attempt to reveal synthesis method, structural and superconducting properties of cuprates materials [45-49]. Apart from the BSCCO scheme, there are several categories of oxides high temperature superconductors like Y–Ba–Cu–O (YBCO family) and T1 originated family T1–Ba–Ca–Cu–O are also vital for technical usages. The YBCO family has exceptional features that lead them towards prominent electromagnetic applications like particle accelerators, power transmission, electronic motors and magnetic levitation gadgets [13]. Also, the T1 family is composed of various phases such as T1 – 2212, T1 – 1223 and T1 – 2223 having T_C above 120 K but have incapability to carry current in wires and electro magnetic tapes. This deficiency is attributed to ineffective control over crystallographic orientation, grain boundaries and micro-cracking [50]. Low density superconductors at higher temperatures have possibility to increase the critical current conductivity (J_C) of superconductors. Low density materials are also known as porous materials having intermediate medium. Sucrose is one of the medium to generate porous surfaces because sucrose does not respond with alumina matrix although burning process [51]. Wu. I. J (2001) studied that the values of J_C for porous YBCO films are greater at least 50 percent as compare to free pores YBCO films even for wide films [13]. The advantage of porous structure over the porous free structure is that it has better contact capability with samples by the application of silver conductive paint which then submerge into the sample materials, creating profound contact instead of pertaining just on the plane.

Y–Ba–Cu–O compound was the earliest having T_c larger than the 77 K and convenient function of superconductivity was significantly accepted after this. Later on, $BiSrCaCu_2O_x$ was the compound with highest T_c [52,53].

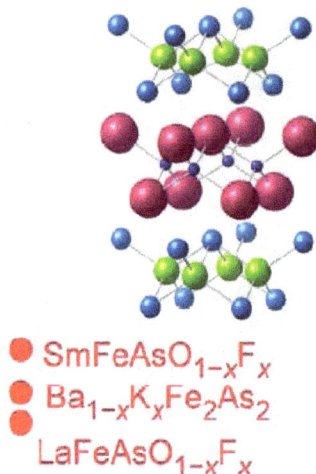

$SmFeAsO_{1-x}F_x$
$Ba_{1-x}K_xFe_2As_2$
$LaFeAsO_{1-x}F_x$

Figure 6 [14]: Structure of iron based superconductors

5.3   Iron based superconductors

Iron based superconductors (IBSCs) had been revealed by Hideo Hosono LaTM*Pn*O where, TM is 3d transition metal and Pn is pnictogen. LaTM*Pn*O have similar crystal structure like LaCuO*Ch* in which Ch is chalcogen ion consisted of a rotating stack of layers of CuCh and LaO. LaCuOCh is belonged to the family of high-gap p-type semiconductor and were discovered by the Hideo Hosono. In 2006 initial IBSC was created for LaFePO; still, T_c was adjusted lower nearly at 4 K [54]. In 2008, higher-T_c compounds were subsequently produced for $LaFeAsO_{1-x}F_x$ where value of x was 0.08 with T_c value 26 K [55]. While in 2007, the LaNiPO was discovered [56]. Again in 2008, MgB_2 was designed at $T_c = 39$ K under the application of high pressure 4 GPa [57].

It was extensively supposed that elements consisting of strong magnetic moments are destructive for appearance of superconductivity due to magnetism which appeared from the stationary magnetic domains. It is due to the fact that both Fe and Ni are the elements having magnetic properties and such creations are also acceptable for condensed matter scientists. Further, progress was done immediately by China, the US and Europe. The quick progress on superconductors was represented in the first international conference in Tokyo on IBSCs which was arranged for the special issue on IBSCs [14] held in 2008. Iron based structure superconductor is given in Figure 6 [14]

6.   Properties of superconductors

Superconductors have vital role for applications in various fields like MRI, NMR and SQUID [58]. Most important parameter of superconductors is the non-appearance of electrical resistance under the transition temperature T_c [59]. Among the different type of superconductors, carbon-based superconductors have significant role [60], which have released the new features in the field of superconductivity. Initial superconductivity was observed at T_c up to 0.55 K in graphite intercalation compound KC8 which is based on carbon materials by Hannay *et al.,* [61]. On the other hand, the study on carbon-based superconductors at high T_c did not stop. Still to the present time, diamond, graphite, carbon nanotubes and fullerene are being investigated for their inherent superconductivity [62-65]. In fact, graphene is the essential part of carbon allotropes consisted of single atom 2 dimension smooth deep layers. Graphene material is widely applicable in industry because of novel properties. Properties of graphene are accredited to high mobility of charge carriers [66], exceptional thermal conductivity [67], visual transparency [68], better mechanical potency [69].

Such interesting characteristics make these materials applicable for supercapacitors, as electrode for spintronic gadgets, LED lighting, solar cells and biosensors [70]. Graphene fundamentally does not act as superconductor because of vanishingly minute electronic density of states at Fermi level [71]. But, have ability to show superconductivity either basically through inserting or substituting the solitary layer of graphene on materials of superconductor [72, 73]. As a result, hybrid structures of graphene with its superconductor arrangements of atoms have been explored for superconductivity. Physical significance for creation of better contacting layers between the superconductor and graphene deposition on upper part is that superconducting coupling associations are created in graphene through proximity with the superconductors. Hence, the graphene having strange band structure has ability to generate super-current. Pairing amongst superconductors and graphene has

Superconductors - Materials and Applications   Materials Research Forum LLC
Materials Research Foundations 132 (2022) 17-48   https://doi.org/10.21741/9781644902110-2

been apprehended and in these joints the super-current has been experimentally observed. Results obtained from the experiments have capability to improve the theoretical viewpoint.

Experimentally, initially the mechanical deposition of graphene was performed on doped silicon substrate and lithographing superconductor [74-76].

Most important was the superconductivity creation at high temperatures in the materials of copper oxide [18]. In case of conventional metals and alloys, created form of BCS superconductivity is interceded via lattice vibrations at which charge carriers make bound couples [77]; under the T_c the condensation also go with s-wave energy distance isotropically at Fermi plane. Inside the structure, T_c was assumed to remain below 30 K [41]. At ambient pressure the T_c is maximumat130 K and strange d-wave energy distance among nodes at Fermi surface, coupling procedure stays undetermined for cuprates. Invention of superconductivity in $MgB_2$by means of unexpectedly elevated T_c = 39 K [39] primarily proposed strange coupling procedure; though, the isotope significance start so as to the superconductivity in these materials is probably phonon mediated [78]. During such conditions, the higher phonon frequencies in MgB_2 materials are major cause of unusual T_c higher value [79]. Detection of superconducting behavior in $LaFeAsO_{1-x}F_x$ (1111) materials, the pnictide complex [55] was found to be astonishing due to the existence of iron. The iron was considered as harmful for the superconductivity process for a longer period of time. The insertion of various rare earth metals in such materials rapidly increase the T_c up to 50 K [80]. This high T_c does not have the capability to knock out the phonon-mediated pairing procedure and orders of magnetism in superconductivity in materials [81] and define such compounds may be consisted of another origin [82]. 1111-family of compounds has maximum T_c and big solo crystals are currently achieved [83]; expanding unit cell of non-superconducting LaFeAsO is illustrated in figure 7 (a). The simpler A-Fe_2As_2 (122) iron-pnictides and FeTe(Se) (11) iron-chalcogenide compounds are given in Figure 7 (b) and Figure 7 (C), respectively.

The metallic AFe_2As_2 compounds in which A = Ca, Ba or Sr are widely examined with the application of pressure in $BaFe_2As_2$ at 29 K T_c [30]. Superconductivity was detected in the arsenic-free iron chalcogenide Fe Se materials where T_c was 8 K, and was enhanced up to 37 K T_c at particular pressure [84]. Although, the insertion of Se in the replacement of Te with the ambient pressure obtained the highest value of 14 K T_c in $FeTe_{0.55}Se_{0.45}$ compounds. Regardless of the distinctions in atomic arrangements of iron-pnictides and the iron-chalcogenides, the band structure of these compounds are extraordinarily alike, with a negligible depiction consisted of an electron band β at the M point and a hole band α centered at the Γ point of the Brillouin zone as demonstrated in figure. 7 [85, 86].

Superconductors - Materials and Applications   Materials Research Forum LLC
Materials Research Foundations 132 (2022) 17-48   https://doi.org/10.21741/9781644902110-2

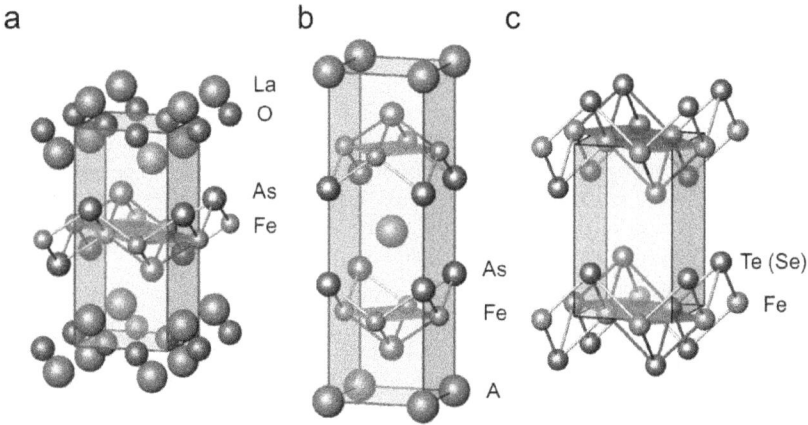

Figure 7 [86]: Iron-pnictides with tetragonally extended unit cells (a) LaFeAsO (b) A Fe₂As₂, where A- alkali earth (c) the iron-chalcogenide FeTe. FeAs sheets are detached by LaO or alkali-earthfilms respectively in (a) and (b).

YBCO is single domain superconductor and nominated as wonderful high-temperature superconductors (HTSC). It is applicable in levitated transportation systems, flywheel energy storage systems, motors, other fields and magnetic bearings [87, 88]. Such usages are dependent on both greater quality and performance YBCO superconductor [89]. To attain the best quality YBCO superconductor, various aspects are needed to be improved like synthesis method, properties and growth [90]. The processing growth of HTSC superconductors is most important now a days and various scientists are working on it. YBCO has heterogeneous crystal structure and the growth is done through molten region for the duration of peritectic reaction amongst particles of Y211 and liquid phase of Ba-Cu-O [91]. Another diffusion model of Y- system based on morphology and distribution of grain growth on the surface composition was designed by Izumi *et al.,* [92]. The macro-segregation of Y211 crystals were synthesised and was revealed that by using optimizing growth factors magnetic flux pinging potential of bulk superconductors can be developed [93]. One domain bulk YBCO superconductors are composed of grains of YBCO with the addition of enclosures like as the particles create highly anisotropic configuration of atoms. Anisotropic grain growth rates of Ra, Rc and c-axis direction are responsible of particle distribution of Y211 and morphology. According to the periodic bond chain (PBC) principle grain growth rate along c-axis direction is 5.4 – 7.3 times slower than the ab-plane [94]. Lo et al. observed that Rc was considerably slower than the Ra by using the morphology of particles in YBCO growth front [95]. The groups of GdBCO samples were grown for 20 hours at various temperatures and were found that growth was lower along the c-axis as compare to the growth rate along ab-plane [96]. Different outcomes were attained for bulk compounds of single domain REBCO [96]. Nakamura et al., investigated that for single domain bulk samples at ΔT greater than 26 K the Ra was lower than the Rc [96, 97]. The growth rate along c-axis is about to achieve via most internal part of compound (Rci). On the other hand, the investigations for a-axis growth rate trend are mostly attained through the upper most surface of compounds (Ras). Few researchers have done work on inner growth rate (Rai) and growth rate of surfaces (Rcs) [98]. Efficient growth rate investigation neither depends on the a-axis, c-axis directions, surface or internal, the foremost important thing is crystal structure growth procedure of YBCO [99].

Upper surface morphology of Pe-Y123 compounds at various super cooling temperatures (ΔT) is shown in the figure 8. Figure 8 exhibits the single domain of YBCCO bulk materials prepared below the 25 K of ΔT, all Pe-Y213 are consisted of epitaxial obtained from NdBCO crystal seed and there is no appearance of any random nucleation. It can also be seen that Pe-Y123 reveals distinctive single domain YBCo bulk morphology below every ΔT. Except dimensions of solitary domain YBCO crystallite developed from NdBCO seed was dissimilar below dissimilar temperature of super cooling, like $\Delta T = 7$ K, single domain region is consisted of relatively slighter dimensions. With the enhancement of super cooling temperature the dimensions of single domain YBCO particles increased. For the development of YBCO superconductor, super cooling temperature is a vital parameter to achieve single domain [100].

Furthermore, the morphology of upper surface of Pe-Y123 family is in square form. The reason is existence of NdBCO crystal seed's c-axis normal to the sample plane. This results in four fold a-axis growth sectors (a-GSs) by means of single domain YBCO bulk lying on apex surface of compounds [99].

Figure 9 [99] shows the Pa-Y123 group bulks with top surface morphology by means of c-axis parallel to upper surface over changed super cooling temperature. The top surface of distinctive YBCO superconductor consisting of single domain is illustrated over variety of ΔT, with no arbitrary nucleation region is detected for all samples. Consequence of super cooling temperature overdevelopment of bulk Pa-Y123 is analogous to outcomes of bulk Pe-Y123 [Figure 8] [99], with the increasing behaviour of super cooling temperature the raising trend is shown.

Pa-Y123 group is consisted of rectangular form, it is due to the upper surface of samples is along the c-axis of seed NdBCO. Such outcomes reveal that surface growth rate of Ras is lower than the Rcs of YBCO bulk when top surfaces of the samples are parallel to c-axis of seed NdBCO [99].

Figure 8 [99]: Upper surface morphology of bulk Pe-Y123 over various super-cooling temperatures ΔT. (From a to j) ΔT is 7 K, 9 K, 11 K, 13 K.

Figure 9 [99]: Upper surface morphology of bulk YBCO over dissimilar supercooling temperatures ΔT. (From a to j) ΔT = 7 K, 9 K, 11 K, 13 K, 15 K, 17 K, 19 K, 21 K, 23 K and 25 K.

Cuprates are consisted of considerable materials after the innovation of high oxide superconductors [101], Bi-based high-temperature superconductors (HTSC). Due to the high transition temperatures of 110 K for Bi-2223 phase, these are applicable in numerous fields because of mechanical characteristics, higher upper critical magnetic field Hc_2 and greater critical current J_c. The BSCCO structures are potential candidates for commercial cables [102], magnets [103], MRI [104], SFCL [105] and magnetic sensors [106]. The properties Bi-2223 mechanism of superconducting are strongly connected with its hole concentrations, pinning ability and phase microstructure [107, 108]. Hence, research importance in BSCCO system is determined principally on recovering the whole characteristics of Bi-2223 phase to improve characteristics of compounds for realistic purposes. The values for J_c may vary from sample to sample depending upon the synthesis states and ratio of doping elements. It is matter of fact that J_c is manipulated through the feeble associated features of the grain boundary amongst the calcined polycrystalline oxide superconductors [109]. The adopted process for the substitution of impurities is applicable to resolve the constraints of J_c and also to estimate the features which influence the superconducting factors significantly. In HTSC the doping percentage of different materials at various states creates the level of crystal imperfections within the compounds. 3-dimensional magnetic elements like cobalt, zinc, nickel, magnesium, iron etc. are potential candidates to be inserted into copper site in copper oxide planes as of their closer ionic radii [110]. Such magnetic insertions repress transition temperatures and also exaggerate value of J_c [111]. Magnetic properties of doping and suppression of superconducting performances are accredited to exclusively limited disorder created due to the insertion of elements. While by the insertion of rare earth metals in these materials the T_c and microstructures show decreasing trends [111]. The flux pinning characteristics of Bi-2223 which is existed in the family of HTSC, by inserting the Nd [112] and Pr [113] showed improving trends. Weak links and flux pinning ability of Bi-2223 phase are applicable in various fields as studied by many scientists [114]. Flux pinning and critical current density can be enhanced by the addition of nano oxide particles in (Bi, Pb)-2223 superconducting phase contributions [115]. J_c, volume pinning force density, U and T_c showed the expanding behaviours by the insertion of 0.2 % Al_2O_3 nanoparticles in polycrystalline HTSC (Bi, Pb)-2223 [116]. Volume fraction ratio and strengthening the (Bi-2223/Bi-2212) phase structure formed by the addition of Sm_2O_3 nanoparticles [117]. The properties of HTSC Bi-2223 can be enhanced by the doping of Co_3O_4 (0.0 – 0.05). When 0.03 % was doped the maximum volume fractions were observed while with the doping of 0.01 % the utmost critical current density was obtained at critical temperature of 102 K [118]. The microstructure, the current density and flux pinning properties of HTSC Bi-2223 systems demonstrated the variable trends by the addition of ZrO_2 [119], SiC [120], MgO [121], Fe_2O_3 [122], MgB_2 [123], Nb_2O_5 [124], Cr_2O_3 and FeS [125] nanoparticles. The addition of doping elements and controlling of grain size may decrease the flux pinning in high temperature superconductors Bi-2223. Further studies are still required to study flux pinning and various mechanisms for classifications in superconducting materials. Few other investigations are considered by substitution of magnetic nanoparticles to illustrate normal-like deficiencies in superconducting matrix, interrelate with the vortices and behave as useful pinning centers inside the high temperature superconductors Bi-2223 materials [113, 126, 127].

Conclusion

The superconductivity was discovered by Kamerlingh Onnes in 1911. The disappearing of electrical resistance below critical temperature (T_c) is known as superconductivity. Two types of superconductors are existed named as type I (soft) and type II (hard) superconductors. Soft superconductors follow the Meissner effect while hard

Superconductors - Materials and Applications
Materials Research Foundations 132 (2022) 17-48

Materials Research Forum LLC
https://doi.org/10.21741/9781644902110-2

superconductors do not obey the Meissner effect. Meissner effect is magnetic property of superconductors. Both type I and type II superconductors have similar superconductivity mechanism and identical thermal properties in normal transition and in zero magnetic field. Superconducting materials are divided into three categories named as; (i) Metal based systems, (ii) Cooper oxides (cuprates) and (iii) Iron based superconductors. The cuprates demonstrate commendable features like strong magnetic field, nil energy failure and current bearing ability. In 2006 the first iron based superconductor was represented for LaFePO; but, the transition temperature was kept nearly at 4 K. Then in 2008 high temperature superconducting materials were exposed at 26 K critical temperature for $LaFeAsO_{1-x}F_x$. Superconductors have various applications in different fields like low temperature magnet applications such as LED lighting, MRI, NMR, SQI, electrode spintronic gadgets, solar cells, supercapacitors and biosensors.

References

[1] D. A. Cardwell, D. S. Ginley, Handbook of superconducting materials, CRC Press 2003. https://doi.org/10.1887/0750308982

[2] A.B. Pippard, The historical context of Josephson's discovery, Superconductor Applications: SQUIDs and Machines, Springer1977, pp. 1-20. https://doi.org/10.1007/978-1-4684-2805-6_1

[3] A. Nikulov, The Law of Entropy Increase and the Meissner Effect, Entropy 24 (2022) 83. https://doi.org/10.3390/e24010083

[4] W. Meissner, R. Ochsenfeld, Ein neuer effekt bei eintritt der supraleitfähigkeit, Naturwissenschaften 21 (1933) 787-788. https://doi.org/10.1007/BF01504252

[5] A. Bleszynski-Jayich, W. Shanks, B. Peaudecerf, E. Ginossar, F. Von Oppen, L. Glazman, J. Harris, Persistent currents in normal metal rings, Science 326 (2009) 272-275. https://doi.org/10.1126/science.1178139

[6] H. Bluhm, N. C. Koshnick, J. A. Bert, M. E. Huber, K. A. Moler, Persistent currents in normal metal rings, Physical Review Letters 102 (2009) 136802. https://doi.org/10.1103/PhysRevLett.102.136802

[7] A. Burlakov, V. Gurtovoï, S. Dubonos, A.V. Nikulov, V. Tulin, Little-parks effect in a system of asymmetric superconducting rings, JETP Letters 86 (2007) 517-521. https://doi.org/10.1134/S0021364007200052

[8] V. Gurtovoi, S. Dubonos, A. Nikulov, N. Osipov, V. Tulin, Dependence of the magnitude and direction of the persistent current on the magnetic flux in superconducting rings, Journal of Experimental and Theoretical Physics 105 (2007) 1157-1173. https://doi.org/10.1134/S1063776107120072

[9] W. Little, R. Parks, Observation of quantum periodicity in the transition temperature of a superconducting cylinder, Physical Review Letters 9 (1962) 9. https://doi.org/10.1103/PhysRevLett.9.9

[10] J. Hirsch, Joule heating in the normal-superconductor phase transition in a magnetic field, Physica C: Superconductivity and its Applications 576 (2020) 1353687. https://doi.org/10.1016/j.physc.2020.1353687

[11] J. R. Waldram, Superconductivity of metals and cuprates, CRC Press 2017. https://doi.org/10.1201/9780203737934

[12] E. Snider, N. Dasenbrock-Gammon, R. McBride, M. Debessai, H. Vindana, K. Vencatasamy, K.V. Lawler, A. Salamat, R.P. Dias, Room-temperature superconductivity in a carbonaceous sulfur hydride, Nature 586 (2020) 373-377. https://doi.org/10.1038/s41586-020-2801-z

[13] E. Nurbaisyatul, H. Azhan, N. Ibrahim, S. Saipuddin, Structural and superconducting properties of low-density Bi (Pb)-2223 superconductor: Effect of Eu2O3 nanoparticles addition, Cryogenics 119 (2021) 103353. https://doi.org/10.1016/j.cryogenics.2021.103353

[14] H. Hosono, A. Yamamoto, H. Hiramatsu, Y. Ma, Recent advances in iron-based superconductors toward applications, Materials today 21 (2018) 278-302. https://doi.org/10.1016/j.mattod.2017.09.006

[15] D. Van Delft, P. Kes, The discovery of superconductivity, Physics Today 63 (2010) 38-43. https://doi.org/10.1063/1.3490499

[16] J. Hirsch, M. Maple, F. Marsiglio, Superconducting materials classes: Introduction and overview, Elsevier, 2015, pp. 1-8. https://doi.org/10.1016/j.physc.2015.03.002

[17] T. Geballe, J. Hulm, Bernd Theodor Matthias, Biographical Memoirs: Volume 70 70 (1996).

[18] J.G. Bednorz, K.A. Müller, Possible highT c superconductivity in the Ba− La− Cu− O system, Zeitschrift für Physik B Condensed Matter 64 (1986) 189-193. https://doi.org/10.1007/BF01303701

[19] H. Takagi, S.-i. Uchida, K. Kitazawa, S. Tanaka, High-Tc superconductivity of La-Ba-Cu oxides. II.-specification of the superconducting phase, Japanese journal of applied physics 26 (1987) L123. https://doi.org/10.1143/JJAP.26.L123

[20] P.A. Lee, N. Nagaosa, X.-G. Wen, Doping a Mott insulator: Physics of high-temperature superconductivity, Reviews of modern physics 78 (2006) 17. https://doi.org/10.1103/RevModPhys.78.17

[21] A.E. Underhill, Molecular metals and superconductors, Journal of Materials Chemistry 2 (1992) 1-11. https://doi.org/10.1039/jm9920200001

[22] H. Akamatu, H. Inokuchi, Y. Matsunaga, Electrical conductivity of the perylene-bromine complex, Nature 173 (1954) 168-169. https://doi.org/10.1038/173168a0

[23] D. Basov, A.V. Chubukov, Manifesto for a higher Tc, Nature Physics 7 (2011) 272-276. https://doi.org/10.1038/nphys1975

Materials Research Forum LLC
https://doi.org/10.21741/9781644902110-2

[24] F. Ronning, T. Klimczuk, E. D. Bauer, H. Volz, J. D. Thompson, Synthesis and properties of CaFe2As2 single crystals, Journal of Physics: Condensed Matter 20 (2008) 322201. https://doi.org/10.1088/0953-8984/20/32/322201

[25] E. Bauer, F. Ronning, B. Scott, J. Thompson, Superconductivity in SrNi2As2 single crystals, Physical Review B 78 (2008) 172504 https://doi.org/10.1103/PhysRevB.78.172504

[26] A. Subedi, D. J. Singh, Density functional study of BaNi2As2: electronic structure, phonons, and electron-phonon superconductivity, Physical Review B 78 (2008) 132511. https://doi.org/10.1103/PhysRevB.78.134514

[27] D. Hirai, F. Von Rohr, R. J. Cava, Emergence of superconductivity in BaNi2(Ge 1−xPx)2 at a structural instability, Physical Review B 86 (2012) 100505. https://doi.org/10.1103/PhysRevB.86.100505

[28] A. S. Sefat, R. Jin, M. A. McGuire, B. C. Sales, D. J. Singh, D. Mandrus, Superconductivity at 22 K in Co-doped BaFe 2 As 2 crystals, Physical review letters 101 (2008) 117004. https://doi.org/10.1103/PhysRevLett.101.117004

[29] L. Shan, J. Gong, Y.-L. Wang, B. Shen, X. Hou, C. Ren, C. Li, H. Yang, H.-H. Wen, S. Li, Evidence of a spin resonance mode in the iron-based superconductor Ba0.6K0.4 Fe2As2 from scanning tunneling spectroscopy, Physical review letters 108 (2012) 227002.

[30] P.L. Alireza, Y. C. Ko, J. Gillett, C.M. Petrone, J.M. Cole, G.G. Lonzarich, S.E. Sebastian, Superconductivity up to 29 K in SrFe2As2 and BaFe2As2 at high pressures, Journal of Physics: Condensed Matter 21 (2008) 012208. https://doi.org/10.1088/0953-8984/21/1/012208

[31] C. Miclea, M. Nicklas, H. Jeevan, D. Kasinathan, Z. Hossain, H. Rosner, P. Gegenwart, C. Geibel, F. Steglich, Evidence for a reentrant superconducting state in EuFe 2 As 2 under pressure, Physical Review B 79 (2009) 212509. https://doi.org/10.1103/PhysRevB.79.212509

[32] W.O. Uhoya, J.M. Montgomery, G.M. Tsoi, Y.K. Vohra, M.A. McGuire, A.S. Sefat, B.C. Sales, S.T. Weir, Phase transition and superconductivity of SrFe2As2 under high pressure, Journal of Physics: Condensed Matter 23 (2011) 122201. https://doi.org/10.1088/0953-8984/23/12/122201

[33] W. Jeitschko, M. Reehuis, Magnetic properties of CaNi2P2 and the corresponding lanthanoid nickel phosphides with ThCr2Si2 type structure, Journal of Physics and Chemistry of Solids 48 (1987) 667-673. https://doi.org/10.1016/0022-3697(87)90157-0

[34] J. An, A.S. Sefat, D.J. Singh, M.-H. Du, Electronic structure and magnetism in BaMn 2 As 2 and BaMn 2 Sb 2, Physical Review B 79 (2009) 075120. https://doi.org/10.1103/PhysRevB.79.075120

[35] Z. Ban, M. Sikirica, The crystal structure of ternary silicides ThM2Si2 (M= Cr, Mn, Fe, Co, Ni and Cu), Acta Crystallographica 18 (1965) 594-599. https://doi.org/10.1107/S0365110X6500141X

[36] W. Jeitschko, W. Hofmann, Ternary alkaline earth and rare earth metal palladium phosphides with ThCr2Si2-and La6Ni6P17-type structures, Journal of the Less Common Metals 95 (1983) 317-322. https://doi.org/10.1016/0022-5088(83)90526-X

[37] J. Blawat, P.W. Swatek, D. Das, D. Kaczorowski, R. Jin, W. Xie, Pd-P antibonding interactions in A Pd 2 P 2 (A= Ca and Sr) superconductors, Physical Review Materials 4 (2020) 014801. https://doi.org/10.1103/PhysRevMaterials.4.014801

[38] A. Schilling, M. Cantoni, J. Guo, H. Ott, Superconductivity above 130 k in the hg-ba-ca-cu-o system, Nature 363 (1993) 56-58. https://doi.org/10.1038/363056a0

[39] J. Nagamatsu, N. Nakagawa, T. Muranaka, Y. Zenitani, J. Akimitsu, Superconductivity at 39 K in magnesium diboride, nature 410 (2001) 63-64. https://doi.org/10.1038/35065039

[40] S.L. Bud'ko, P.C. Canfield, Superconductivity of magnesium diboride, Physica C: Superconductivity and its Applications 514 (2015) 142-151. https://doi.org/10.1016/j.physc.2015.02.024

[41] W. McMillan, Transition temperature of strong-coupled superconductors, Physical Review 167 (1968) 331. https://doi.org/10.1103/PhysRev.167.331

[42] W.E. Pickett, Design for a room-temperature superconductor, Journal of Superconductivity and Novel Magnetism 19 (2006) 291-297. https://doi.org/10.1007/s10948-006-0164-9

[43] H. Fallah-Arani, S. Baghshahi, A. Sedghi, Impact of functionalized SiC nano-whisker on the flux pinning ability and superconductor features of Bi-2223 ceramics, Ceramics International 47 (2021) 3706-3712. https://doi.org/10.1016/j.ceramint.2020.09.226

[44] M. Anis-ur-Rehman, M. Mubeen, Synthesis and enhancement of current density in cerium doped Bi (Pb) Sr (Ba)-2 2 2 3 high Tc superconductor, Synthetic Metals 162 (2012) 1769-1774. https://doi.org/10.1016/j.synthmet.2012.03.006

[45] S. Chu, M. McHenry, Critical current density in high-Tc Bi-2223 single crystals using AC and DC magnetic measurements, Physica C: Superconductivity 337 (2000) 229-233. https://doi.org/10.1016/S0921-4534(00)00107-6

[46] I. Hamadneh, A. Agil, A. Yahya, S. Halim, Superconducting properties of bulk Bi1. 6Pb0. 4Sr2Ca2− xCdxCu3O10 system prepared via conventional solid state and coprecipitation methods, Physica C: Superconductivity and its applications 463 (2007) 207-210. https://doi.org/10.1016/j.physc.2007.03.445

[47] J. Hawa, H. Azhan, S. Yahya, K. Azman, H. Hidayah, A. Norazidah, The effect of Eu substitution onto Ca site in Bi (Pb)-2223 superconductor via co-precipitation

method, Journal of superconductivity and novel magnetism 26 (2013) 979-983. https://doi.org/10.1007/s10948-012-2043-x

[48] J.S. Hawa, A. Hashim, S. Yahya, A. Kasim, H.N. Hidayah, A.W. Norazidah, Properties of Rare-Earth Substitution in Bi (Pb)-2223 Superconductor Prepared by Coprecipitation Method, Advanced Materials Research, Trans Tech Publ, 2014, pp. 83-86. https://doi.org/10.4028/www.scientific.net/AMR.895.83

[49] N.H. Mohammed, R. Awad, A.I. Abou-Aly, I.H. Ibrahim, M.S. Hassan, Optimizing the preparation conditions of Bi-2223 superconducting phase using PbO and PbO 2, (2012). https://doi.org/10.4236/msa.2012.34033

[50] M. Siegal, E. Venturini, B. Morosin, T. Aselage, Synthesis and properties of Tl-Ba-Ca-Cu-O superconductors, Journal of materials research 12 (1997) 2825-2854. https://doi.org/10.1557/JMR.1997.0378

[51] Z. Silveira, R. Nicoletti, C. Fortulan, B. Purquerio, Ceramic matrices applied to aerostatic porous journal bearings: material characterization and bearing modeling, Cerâmica 56 (2010) 201-211. https://doi.org/10.1590/S0366-69132010000200016

[52] M.-K. Wu, J.R. Ashburn, C. Torng, P.-H. Hor, R.L. Meng, L. Gao, Z.J. Huang, Y. Wang, a. Chu, Superconductivity at 93 K in a new mixed-phase Y-Ba-Cu-O compound system at ambient pressure, Physical review letters 58 (1987) 908. https://doi.org/10.1103/PhysRevLett.58.908

[53] H. Maeda, Y. Tanaka, M. Fukutomi, T. Asano, A new high-Tc oxide superconductor without a rare earth element, Japanese Journal of Applied Physics 27 (1988) L209. https://doi.org/10.1143/JJAP.27.L209

[54] Y. Kamihara, H. Hiramatsu, M. Hirano, R. Kawamura, H. Yanagi, T. Kamiya, H. Hosono, Iron-based layered superconductor: LaOFeP, Journal of the American Chemical Society 128 (2006) 10012-10013. https://doi.org/10.1021/ja063355c

[55] Y. Kamihara, T. Watanabe, M. Hirano, H. Hosono, Iron-based layered superconductor La [O1-x F x] FeAs (x= 0.05− 0.12) with T c= 26 K, Journal of the American Chemical Society 130 (2008) 3296-3297. https://doi.org/10.1021/ja800073m

[56] T. Watanabe, H. Yanagi, T. Kamiya, Y. Kamihara, H. Hiramatsu, M. Hirano, H. Hosono, Nickel-based oxyphosphide superconductor with a layered crystal structure, LaNiOP, Inorganic Chemistry 46 (2007) 7719-7721. https://doi.org/10.1021/ic701200e

[57] H. Takahashi, K. Igawa, K. Arii, Y. Kamihara, M. Hirano, H. Hosono, Superconductivity at 43 K in an iron-based layered compound LaO1-xFxFeAs, nature 453 (2008) 376-378. https://doi.org/10.1038/nature06972

[58] Y. Liu, H. Liang, Z. Xu, J. Xi, G. Chen, W. Gao, M. Xue, C. Gao, Superconducting continuous graphene fibers via calcium intercalation, ACS nano 11 (2017) 4301-4306. https://doi.org/10.1021/acsnano.7b01491

[59] P.B. Atienza, Superconductivity in Graphene and Carbon Nanotubes: Proximity effect and nonlocal transport, Springer Science & Business Media2013.

[60] J. Haruyama, Carbon-based Superconductors: Towards High-Tc Superconductivity, CRC Press2014. https://doi.org/10.1201/b15672

[61] N. Hannay, T. Geballe, B. Matthias, K. Andres, P. Schmidt, D. MacNair, Superconductivity in graphitic compounds, Physical Review Letters 14 (1965) 225. https://doi.org/10.1103/PhysRevLett.14.225

[62] M.S. Dresselhaus, G. Dresselhaus, Intercalation compounds of graphite, Advances in physics 51 (2002) 1-186. https://doi.org/10.1080/00018730110113644

[63] E. Ekimov, V. Sidorov, E. Bauer, N. Mel'Nik, N. Curro, J. Thompson, S. Stishov, Superconductivity in diamond, nature 428 (2004) 542-545. https://doi.org/10.1038/nature02449

[64] A. Hebard, M. Rosseinky, R. Haddon, D. Murphy, S. Glarum, T. Palstra, A. Ramirez, A. Karton, Potassium-doped C60, Nature 350 (1991) 600-601. https://doi.org/10.1038/350600a0

[65] Z. Tang, L. Zhang, N. Wang, X. Zhang, G. Wen, G. Li, J. Wang, C.T. Chan, P. Sheng, Superconductivity in 4 angstrom single-walled carbon nanotubes, Science 292 (2001) 2462-2465. https://doi.org/10.1126/science.1060470

[66] K.I. Bolotin, K.J. Sikes, Z. Jiang, M. Klima, G. Fudenberg, J. Hone, P. Kim, H. Stormer, Ultrahigh electron mobility in suspended graphene, Solid state communications 146 (2008) 351-355. https://doi.org/10.1016/j.ssc.2008.02.024

[67] A.A. Balandin, S. Ghosh, W. Bao, I. Calizo, D. Teweldebrhan, F. Miao, C.N. Lau, Superior thermal conductivity of single-layer graphene, Nano letters 8 (2008) 902-907. https://doi.org/10.1021/nl0731872

[68] R.R. Nair, P. Blake, A.N. Grigorenko, K.S. Novoselov, T.J. Booth, T. Stauber, N.M. Peres, A.K. Geim, Fine structure constant defines visual transparency of graphene, Science 320 (2008) 1308-1308. https://doi.org/10.1126/science.1156965

[69] C. Lee, X. Wei, J.W. Kysar, J. Hone, Measurement of the elastic properties and intrinsic strength of monolayer graphene, science 321 (2008) 385-388. https://doi.org/10.1126/science.1157996

[70] A.C. Ferrari, F. Bonaccorso, V. Fal'Ko, K.S. Novoselov, S. Roche, P. Bøggild, S. Borini, F.H. Koppens, V. Palermo, N. Pugno, Science and technology roadmap for graphene, related two-dimensional crystals, and hybrid systems, Nanoscale 7 (2015) 4598-4810. https://doi.org/10.1039/C4NR01600A

[71] A.P. Tiwari, S. Shin, E. Hwang, S.-G. Jung, T. Park, H. Lee, Superconductivity at 7.4 K in few layer graphene by Li-intercalation, Journal of Physics: Condensed Matter 29 (2017) 445701. https://doi.org/10.1088/1361-648X/aa88fb

[72] J. Linder, A.M. Black-Schaffer, T. Yokoyama, S. Doniach, A. Sudbø, Josephson current in graphene: Role of unconventional pairing symmetries, Physical Review B 80 (2009) 094522. https://doi.org/10.1103/PhysRevB.80.094522

[73] B. Uchoa, A.C. Neto, Superconducting states of pure and doped graphene, Physical review letters 98 (2007) 146801. https://doi.org/10.1103/PhysRevLett.98.146801

[74] H.B. Heersche, P. Jarillo-Herrero, J.B. Oostinga, L.M. Vandersypen, A.F. Morpurgo, Bipolar supercurrent in graphene, Nature 446 (2007) 56-59. https://doi.org/10.1038/nature05555

[75] F. Miao, S. Wijeratne, Y. Zhang, U. Coskun, W. Bao, C. Lau, Phase-coherent transport in graphene quantum billiards, science 317 (2007) 1530-1533. https://doi.org/10.1126/science.1144359

[76] G. Xu, B. Wu, J. Cao, Alternating current Josephson effect in superconductor-graphene-superconductor junctions, Journal of Applied Physics 109 (2011) 083704. https://doi.org/10.1063/1.3573501

[77] J. Bardeen, L.N. Cooper, J.R. Schrieffer, Theory of superconductivity, Physical review 108 (1957) 1175. https://doi.org/10.1103/PhysRev.108.1175

[78] P. Canfield, D. Finnemore, S. Bud'Ko, J. Ostenson, G. Lapertot, C. Cunningham, C. Petrovic, Superconductivity in dense MgB 2 wires, Physical Review Letters 86 (2001) 2423. https://doi.org/10.1103/PhysRevLett.86.2423

[79] J. Kortus, I. Mazin, K.D. Belashchenko, V.P. Antropov, L. Boyer, Superconductivity of metallic boron in MgB 2, Physical Review Letters 86 (2001) 4656. https://doi.org/10.1103/PhysRevLett.86.4656

[80] Z.-A. Ren, J. Yang, W. Lu, W. Yi, X.-L. Shen, Z.-C. Li, G.-C. Che, X.-L. Dong, L.-L. Sun, F. Zhou, Superconductivity in the iron-based F-doped layered quaternary compound Nd [O1− x Fx] FeAs, EPL (Europhysics Letters) 82 (2008) 57002. https://doi.org/10.1209/0295-5075/82/57002

[81] C. de La Cruz, Q. Huang, J. Lynn, J. Li, W.R. Ii, J.L. Zarestky, H. Mook, G. Chen, J. Luo, N. Wang, Magnetic order close to superconductivity in the iron-based layered lao1-xf x feas systems, nature 453 (2008) 899-902. https://doi.org/10.1038/nature07057

[82] L. Boeri, O. Dolgov, A.A. Golubov, Is LaFeAsO 1− x F x an electron-phonon superconductor?, Physical Review Letters 101 (2008) 026403. https://doi.org/10.1103/PhysRevLett.101.026403

[83] J.-Q. Yan, S. Nandi, J.L. Zarestky, W. Tian, A. Kreyssig, B. Jensen, A. Kracher, K.W. Dennis, R.J. McQueeney, A.I. Goldman, Flux growth at ambient pressure of

millimeter-sized single crystals of LaFeAsO, LaFeAsO 1− x F x, and LaFe 1− x Co x
AsO, Applied Physics Letters 95 (2009) 222504. https://doi.org/10.1063/1.3268435

[84] F.-C. Hsu, J.-Y. Luo, K.-W. Yeh, T.-K. Chen, T.-W. Huang, P.M. Wu, Y.-C. Lee,
Y.-L. Huang, Y.-Y. Chu, D.-C. Yan, Superconductivity in the PbO-type structure α-
FeSe, Proceedings of the National Academy of Sciences 105 (2008) 14262-14264.
https://doi.org/10.1073/pnas.0807325105

[85] S. Raghu, X.-L. Qi, C.-X. Liu, D. Scalapino, S.-C. Zhang, Minimal two-band model
of the superconducting iron oxypnictides, Physical Review B 77 (2008) 220503.
https://doi.org/10.1103/PhysRevB.77.220503

[86] C.C. Homes, A. Akrap, J. Wen, Z. Xu, Z.W. Lin, Q. Li, G. Gu, Optical properties of
the iron-chalcogenide superconductor FeTe0. 55Se0. 45, Journal of Physics and
Chemistry of Solids 72 (2011) 505-510. https://doi.org/10.1016/j.jpcs.2010.10.014

[87] B. Oswald, K. Best, M. Setzer, M. Söll, W. Gawalek, A. Gutt, L. Kovalev, G.
Krabbes, L. Fisher, H. Freyhardt, Reluctance motors with bulk HTS material,
Superconductor Science and Technology 18 (2004) S24. https://doi.org/10.1088/0953-
2048/18/2/006

[88] F. Werfel, U. Floegel-Delor, R. Rothfeld, B. Goebel, D. Wippich, T. Riedel,
Modelling and construction of a compact 500 kg HTS magnetic bearing,
Superconductor Science and Technology 18 (2004) S19. https://doi.org/10.1088/0953-
2048/18/2/005

[89] R.-P. Sawh, R. Weinstein, K. Carpenter, D. Parks, K. Davey, Production run of 2 cm
diameter YBCO trapped field magnets with surface field of 2 T at 77 K,
Superconductor Science and Technology 26 (2013) 105014.
https://doi.org/10.1088/0953-2048/26/10/105014

[90] Y.-X. Guo, W.-M. Yang, J.-W. Li, L.-P. Guo, L.-P. Chen, Q. Li, Effects of vertical
temperature gradient on the growth morphology and properties of single domain
YBCO bulks fabricated by a new modified TSIG technique, Crystal Growth & Design
15 (2015) 1771-1775. https://doi.org/10.1021/cg501817z

[91] C.A. Bateman, L. Zhang, H.M. Chan, M.P. Harmer, Mechanism for the peritectic
reaction and growth of aligned grains in YBa2Cu3O6+ x, Journal of the American
Ceramic Society 75 (1992) 1281-1283. https://doi.org/10.1111/j.1151-
2916.1992.tb05572.x

[92] P. Barua, V. Srinivas, S. Dhabal, T. Ghosh, X-ray photoelectron spectroscopic
studies on nanoquasicrystalline powders of Al70Cu20Fe10 obtained by mechanical
alloying, Journal of materials research 17 (2002) 1892-1895.
https://doi.org/10.1557/JMR.2002.0280

[93] A. Endo, H. Chauhan, T. Egi, Y. Shiohara, Macrosegregation of Y {sub 2} Ba {sub
1} Cu {sub 1} O {sub 5} particles in Y {sub 1} Ba {sub 2} Cu {sub 3} O {sub 7

{minus}{delta}} crystals grown by an undercooling method, Journal of Materials Research 11 (1996). https://doi.org/10.1557/JMR.1996.0096

[94] B.N. Sun, P. Hartman, C. Woensdregt, H. Schmid, Structural morphology of YBa2Cu3O7-x, Journal of crystal growth 100 (1990) 605-614. https://doi.org/10.1016/0022-0248(90)90259-N

[95] W. Lo, D. Cardwell, J. Chow, Anisotropic growth morphology and platelet formation in large grain Y-Ba-Cu-O grown by seeded peritectic solidification, Journal of materials research 13 (1998) 1141-1146. https://doi.org/10.1557/JMR.1998.0162

[96] G.Z. Li, W.M. Yang, Y.L. Tang, J. Ma, Growth of single-grain GdBa2Cu3O7-x superconductors by top seeded infiltration and growth technique, Crystal Research and Technology: Journal of Experimental and Industrial Crystallography 45 (2010) 219-225. https://doi.org/10.1002/crat.200900365

[97] Y. Nakamura, A. Endo, Y. Shiohara, The relation between the undercooling and the growth rate of YBa2Cu3O6+ x superconductive oxide, Journal of materials research 11 (1996) 1094-1100. https://doi.org/10.1557/JMR.1996.0139

[98] M. Radusovska, P. Diko, S. Piovarci, S. Park, B. Jun, C. Kim, Microstructure and trapped field of YBCO bulk single-grain superconductors prepared by interior seeding, Superconductor Science and Technology 30 (2017) 105013. https://doi.org/10.1088/1361-6668/aa8648

[99] T.-T. Wu, H.-Y. Zhang, W.-M. Yang, Y.-L. Cui, A. Abulaiti, Effect of seed orientations on the crystallization characteristics and the magnetic levitation properties of single domain YBCO bulk superconductors, Journal of Alloys and Compounds 883 (2021) 160788. https://doi.org/10.1016/j.jallcom.2021.160788

[100] W. Yang, X. Guo, F. Wan, G. Li, Real-time observation and analysis of single-domain YBCO bulk superconductor by TSIG process, Crystal growth & design 11 (2011) 3056-3059. https://doi.org/10.1021/cg2003222

[101] C. Lee, W. Yang, R.G. Parr, Development of the Colle-Salvetti correlation-energy formula into a functional of the electron density, Physical review B 37 (1988) 785. https://doi.org/10.1103/PhysRevB.37.785

[102] T. Masuda, T. Kato, H. Yumura, M. Watanabe, Y. Ashibe, K. Ohkura, C. Suzawa, M. Hirose, S. Isojima, K. Matsuo, Verification tests of a 66 kV HTSC cable system for practical use (first cooling tests), Physica C: Superconductivity 378 (2002) 1174-1180. https://doi.org/10.1016/S0921-4534(02)01750-1

[103] T. Kurusu, M. Ono, S. Hanai, M. Kyoto, H. Takigami, H. Takano, K. Watanabe, S. Awaji, K. Koyama, G. Nishijima, A cryocooler-cooled 19 T superconducting magnet with 52 mm room temperature bore, IEEE transactions on applied superconductivity 14 (2004) 393-396. https://doi.org/10.1109/TASC.2004.829679

Materials Research Forum LLC
https://doi.org/10.21741/9781644902110-2

[104] M. Cheng, B. Yan, K. Lee, Q. Ma, E. Yang, A high temperature superconductor tape RF receiver coil for a low field magnetic resonance imaging system, Superconductor Science and Technology 18 (2005) 1100. https://doi.org/10.1088/0953-2048/18/8/013

[105] K. Suzuki, J. Baba, T. Nitta, Conceptual design of an SFCL by use of BSCCO wire, Journal of Physics: Conference Series, IOP Publishing, 2008, pp. 012293. https://doi.org/10.1088/1742-6596/97/1/012293

[106] B. Albiss, Thick films of superconducting YBCO as magnetic sensors, Superconductor Science and Technology 18 (2005) 1222. https://doi.org/10.1088/0953-2048/18/9/014

[107] R. Aloysius, P. Guruswamy, U. Syamaprasad, Enhanced critical current density in (Bi, Pb)-2223 superconductor by Nd addition in low percentages, Physica C: Superconductivity 426 (2005) 556-562. https://doi.org/10.1016/j.physc.2005.05.017

[108] C. Terzioglu, M. Yilmazlar, O. Ozturk, E. Yanmaz, Structural and physical properties of Sm-doped Bi1. 6Pb0. 4Sr2Ca2− xSmxCu3Oy superconductors, Physica C: Superconductivity and its applications 423 (2005) 119-126. https://doi.org/10.1016/j.physc.2005.04.008

[109] J. Ekin, A.I. Braginski, A. Panson, M. Janocko, D. Capone, N. Zaluzec, B. Flandermeyer, O. De Lima, M. Hong, J. Kwo, Evidence for weak link and anisotropy limitations on the transport critical current in bulk polycrystalline Y1Ba2Cu3O x, Journal of applied physics 62 (1987) 4821-4828. https://doi.org/10.1063/1.338985

[110] S. Sinha, S. Gadkari, S. Sabharwal, L. Gupta, M. Gupta, Bulk synthesis of (BiPb) 2Sr2Ca2Cu3Ox superconductor, Physica C: Superconductivity 185 (1991) 499-500. https://doi.org/10.1016/0921-4534(91)92052-D

[111] A. Mamalis, S. Ovchinnikov, M. Petrov, D. Balaev, K. Shaihutdinov, D. Gohfeld, S. Kharlamova, I. Vottea, Composite materials on high-Tc superconductors and BaPbO3, Ag basis, Physica C: Superconductivity and its applications 364 (2001) 174-177. https://doi.org/10.1016/S0921-4534(01)00749-3

[112] X. Cai, A. Gurevich, D. Larbalestier, R. Kelley, M. Onellion, H. Berger, G. Margaritondo, Static and dynamic mechanisms of the anomalous field dependence of magnetization in Bi-Sr-Ca-Cu-O and Bi-Pb-Sr-Ca-Cu-O single crystals, Physical Review B 50 (1994) 16774. https://doi.org/10.1103/PhysRevB.50.16774

[113] M. Shalaby, M.H. Hamed, N. Yousif, H. Hashem, The impact of the addition of Bi2Te3 nanoparticles on the structural and the magnetic properties of the Bi-2223 high-Tc superconductor, Ceramics international 47 (2021) 25236-25248. https://doi.org/10.1016/j.ceramint.2021.05.244

[114] Z. Jia, H. Tang, Z. Yang, Y. Xing, Y. Wang, G. Qiao, Effects of nano-ZrO2 particles on the superconductivity of Pb-doped BSCCO, Physica C: Superconductivity 337 (2000) 130-132. https://doi.org/10.1016/S0921-4534(00)00072-1

[115] E. Guilmeau, B. Andrzejewski, J. Noudem, The effect of MgO addition on the formation and the superconducting properties of the Bi2223 phase, Physica C: Superconductivity 387 (2003) 382-390. https://doi.org/10.1016/S0921-4534(02)02360-2

[116] A. Ghattas, M. Zouaoui, M. Annabi, A. Madani, F.B. Azzouz, M.B. Salem, Enhancement of superconductivity properties in nano ZrO2 particles added Bi1. 8Pb0. 4Sr2Ca2Cu3Ox ceramics, Journal of Physics: Conference Series, IOP Publishing, 2008, pp. 012179. https://doi.org/10.1088/1742-6596/97/1/012179

[117] H. Baqiah, S. Halim, M. Adam, S. Chen, S. Ravandi, M. Faisal, M. Kamarulzaman, M. Hanif, The Effect of Magnetic Nanoparticle Addition on the Superconducting Properties Bi1. 6Pb0. 4Sr2Ca2Cu3Oδ Superconductors, Solid State Science and Technology 17 (2009) 81-88.

[118] S. Zhang, X. Ma, B. Shao, L. Cui, G. Liu, H. Zheng, X. Liu, J. Feng, C. Li, P. Zhang, Fabrication of multifilamentary powder in tube superconducting tapes of Bi-2223 with Sr deficient starting composition, Cryogenics 114 (2021) 103245. https://doi.org/10.1016/j.cryogenics.2020.103245

[119] Y. Guo, Y. Tanaka, T. Kuroda, S. Dou, Z. Yang, Addition of nanometer SiC in the silver-sheathed Bi2223 superconducting tapes, Physica C: Superconductivity 311 (1999) 65-74. https://doi.org/10.1016/S0921-4534(98)00625-X

[120] W. Wei, J. Schwartz, K. Goretta, U. Balachandran, A. Bhargava, Effects of nanosize MgO additions to bulk Bi2. 1Sr1. 7CaCu2Ox, Physica C: Superconductivity 298 (1998) 279-288. https://doi.org/10.1016/S0921-4534(97)01889-3

[121] K. Wei, R. Abd-Shukor, Superconducting and transport properties of (Bi-Pb)-Sr-Ca-Cu-O with nano-Cr2O3 additions, Journal of Electronic Materials 36 (2007) 1648-1651. https://doi.org/10.1007/s11664-007-0287-1

[122] M. Annabi, A. M'chirgui, F.B. Azzouz, M. Zouaoui, M.B. Salem, Addition of nanometer Al2O3 during the final processing of (Bi, Pb)-2223 superconductors, Physica C: Superconductivity 405 (2004) 25-33. https://doi.org/10.1016/j.physc.2004.01.012

[123] H. Sözeri, N. Ghazanfari, H. Özkan, A. Kilic, Enhancement in the high-Tc phase of BSCCO superconductors by Nb addition, Superconductor Science and Technology 20 (2007) 522. https://doi.org/10.1088/0953-2048/20/6/007

[124] K. Habanjar, F.E.H. Hassan, R. Awad, Physical and dielectric properties of (Bi, Pb)-2223 superconducting samples added with BaFe12O19 nanoparticles, Chemical Physics Letters 757 (2020) 137880. https://doi.org/10.1016/j.cplett.2020.137880

[125] M. Masnita, R. Abd-Shukor, Iron sulfide effects on AC susceptibility and electrical properties of Bi1. 6Pb0. 4Sr2CaCu2O8 superconductor, Results in Physics 17 (2020) 103177. https://doi.org/10.1016/j.rinp.2020.103177

Superconductors - Materials and Applications   Materials Research Forum LLC
Materials Research Foundations 132 (2022) 17-48   https://doi.org/10.21741/9781644902110-2

[126] D. Brochier, P. Cardinne, M. Renard, Inclusions ferromagnétiques dans des supraconducteurs de deuxième espèce, Journal de Physique 29 (1968) 953-956. https://doi.org/10.1051/jphys:019680029010095300

[127] P. Togulev, V. Bazarov, I. Khaïbullin, N. Suleïmanov, Reinforcement of pinning by surface magnetic microparticles in high-Tc superconductors, Low Temperature Physics 28 (2002) 250-253. https://doi.org/10.1063/1.1477357

Superconductors - Materials and Applications
Materials Research Foundations 132 (2022) 49-78

Materials Research Forum LLC
https://doi.org/10.21741/9781644902110-3

Chapter 3

Fundamentals and Properties of Superconductors

M. Rizwan[1*], A. Ayub[2], S. Fatima[2], I. Ilyas[2], A. Usman[2], A. Shoukat[3]

[1]School of Physical Sciences, University of the Punjab, Lahore, Pakistan

[2]Department of Physics, University of the Punjab, Lahore, Pakistan

[3]Department of Physics, University of Gujrat, Gujrat, Pakistan

*rizwan.sps@pu.edu.pk

Abstract

Superconductors are evolutionary materials that have applications in energy harvesting, magnet technology, power generation and transmission, maglev transportation and many more due to its outstanding properties such as zero resistance, perfect conductivity and diamagnetism and Meissner effect. These potential applications are the key behind the boost in superconductor field and research. Properties and exploration of superconductors in depth is very crucial to decide its future and potential in many technologies. Superconductors are a fast field and thus are classified in many categories such as type I and II semiconductors, high and low T_c superconductors based on coherence length and critical temperature respectively.

Keywords

High Temperature Superconductors, Superconductivity, Landau Theory, Supercurrents, Meissner Effect, Josephson Junction

Contents

Superconductors - Materials and Applications   Materials Research Forum LLC
Materials Research Foundations 132 (2022) 49-78   https://doi.org/10.21741/9781644902110-3

1.   Introduction

Superconductors and the phenomenon of superconductivity still hold their charm ever since their discovery in 1911 by Heike Kamerlingh Onnes [1]. Superconductors are materials that exhibit zero resistance below a certain temperature and thus give maximum current. The ultimate dream is to have current flow with no resistance. The ratio of resistivity between metallic and superconducting states has exceeded the value of 10^{14}, but the concept of superconductivity is more than a zero-resistance state and perfect conductor state. In 1933, Meissner and Ochsenfeld discovered an important property of superconductors, flux was expelled from superconductors and this was called Meisner effect. It was independent of any experimental history and superconductivity was rendered as the absolute thermodynamic state of a matter [2]. Expulsion of magnetic field necessitates zero resistance, this perfect diamagnetism condition indicates that superconducting state is a true thermodynamic state as material undergoes from normal metallic state to thermodynamic state, it undergoes a thermodynamic phase transition [3].

Cooling techniques improved with time and gave access to lower temperatures and thus superconductivity was established as low temperature instability for almost all metallic systems. BCS (Bardeen, Cooper, Schrieffer) theory in 1957 [4] further improved the superconductivity explanation [5]. The first ever superconductor was mercury which showed superconducting behavior below 4.2K. Most superconductors were low temperature superconductors in that time. In metals niobium is the highest temperature semiconductor material with critical temperature T_C of 9.25K. Bednorz and Muller discovered High temperature superconductors in 1986, with present high critical temperature of 135K [6].

The fundamental difference between metals, insulators, conductors and the class of superconductors is kept upon their resistivity. The conductivity of a material is a direct property of electrical resistivity; the resistivity is low for good conductors, high for bad conductors such as insulators. In conductors there are electrons which are in continuous motion due to thermal motion, this motion is exalted at low temperatures, and this resistance reduces in low temperature region but still remains finite [7]. In comparison, in superconductors the resistivity becomes extremely low at critical temperature. This behavior of resistivity in metals and superconductors is shown in Fig.1 [1].

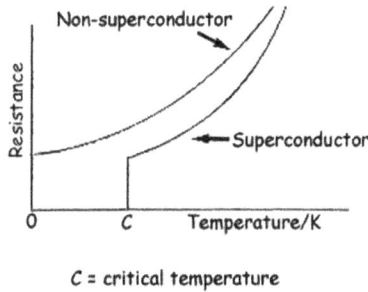

C = critical temperature

Figure .1. *Behavior of electrical resistivity with temperature for metals and superconductors [8].*

Electrons bind and give isotropic cooper pairs via interaction with ion presents in its surrounding. Over last four decades, researchers have tried to look for magnetic superconductors that have unconventional pairing. One of such unconventional superconductor was reported in 1994 by Maeno and company [9], strontium ruthenate that showed superconductivity near 1K. The comprehension of such unconventional superconductors requires total understanding of the pairing interaction and the order parameter. Order parameter tells about interaction and macroscopic properties such as anisotropy [10]. The most exciting advancement in the field of superconductivity happened in late 1980s, and that was the discovery of the phenomenon in layered cooperates. This superconductivity was seen at a high temperature of 30K and gave a way to get a gist of cuperate superconductivity [6]. The discovery of high temperature cooperate superconductors set a new era in the field of conductivity and prompted research in this field. Within few months of it discover may High T_c superconductors came to light which easily exceeded liquid nitrogen temperatures thus further expanding the scope and future of the field and its applications.

For a long time cuperates was considered to be the only class of high temperature superconductors, but in 2001, Akimistu [11] and company discovered another class of superconductors called MgB_2 with a T_c of 39K. This value of T_c was far lower than the highest value of T_c for cuperates, but it exceeds for normal ones. In 2006, superconductivity was observed in iron based materials and T_c was 56K which is still far below than for cooperates [12]. H_2S was turned into superconductor by Eremets [13] in 2015 with T_c >200K. This discovery suggested that hydrogen-based materials act as potential contenders for room temperature superconductors, which is the most studied class of superconductors in recent times. Recently superconductivity at room temperature was observed in carbonaceous sulfur hydride with T_c up to 287.7 K under 267 GPa pressure, but there are still concerns about the type of superconductivity [14]. Even though room temperature superconductors appear fancier superconductors, but their applications at such higher temperatures is difficult to realize than at low temperatures [15]. The critical temperature for superconductors has progressed a lot ever since the discovery of superconductors [16]. From 4.2K, the research is now focused towards room temperature (300K) [17]. The evolution of critical temperature, T_c, over the years is shown in Fig.2 [1]. The list of superconductors discerned till yet and their critical temperatures are given in Table 1.

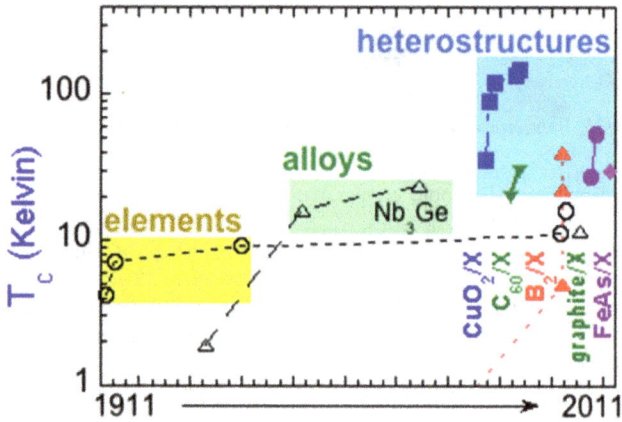

Figure 2. *Explanation of critical temperature (T_c) of superconductors over the years [18]*

Table 1. *List of superconductors and their reported Critical temperature,* T_c

Substance	TC (K)	Ref.	Substance	TC (K)	Ref.e
Al	1.20	[19]	C_3Na	2.3–3.8	[20]
Bi	5.3×10^{-4}	[19]	C_2Na	5.0	[20]
Cd	0.52	[21]	C_8Rb	0.025	[22]
Diamond: B	11.4	[23]	C_6Sr	1.65	[22]
Ga	1.083	[21]	C_6Yb	6.5	[22]
Hf	0.165	[21]	$C_{60}Cs_2Rb$	33	[24]
α-Hg	4.15	[21]	$C_{60}K_3$	19.8	[22]
β-Hg	3.95	[21]	$C_{60}RbX$	28	[25]
In	3.4	[21]	FeB_4	2.9	[21]
Ir	0.14	[21]	InN	3	[26]
α-La	4.9	[21]	In_2O_3	3.3	[27]
β-La	6.3	[21]	LaB_6	0.45	[28]
Li	4×10^{-4}	[9]	MgB_2	39	[29]
Mo	0.92	[2]	Nb_3Al	18	[21]
Nb	9.26	[21]	$NbC_{1-x}N_x$	17.8	[30]
Os	0.65	[21]	Nb_3Ge	23.2	[31]
Pa	1.4	[32]	NbO	1.38	[33]
Pb	7.19	[21]	NbN	16	[21]
Re	2.4	[34]	Nb_3Sn	18.3	[35]
Rh	3.25×10^{-4}	[12]	NbTi	10	[21]
Ru	0.49	[34]	SiC:B	1.4	[36]
Si:B	0.4	[13]	SiC:Al	1.5	18]
Sn	3.72	[21]	TiN	5.6	[37]
Ta	4.48	[34]	V_3Si	17	[38]
Tc	7.46–11.2	[34]	YB_6	8.4	[39]
α-Th	1.37	[34]	ZrN	10	[40]
Ti	0.39	[34]	ZrB_{12}	6.0	[41]
α-U	0.68	[32]	EuBCO	93	[42]
β-U	1.8	[32]	GdBCO	91	[43]
V	5.03	[34]	α-W	0.015	[32]
Zn	0.855	[34]	β-W	1–4	[44]
Zr	0.55	[34]	Ba_8Si_{46}	8.07	[22]
C_6Ca	11.5	[22]	$C_6Li_3Ca_2$	11.15	[22]
C_8K	0.14	[22]	C_8KHg	1.4	[22]
C_6K	1.5	[22]	C_3K	3.0	[20]
C_3Li	<0.35	[s20]	C_2Li	1.9	[17]

Superconductors - Materials and Applications   Materials Research Forum LLC
Materials Research Foundations 132 (2022) 49-78   https://doi.org/10.21741/9781644902110-3

The motivation towards high temperature superconductors is because of their properties and potential to be superior alterntaive in many applications. Superconductors are essential for appplications which require high current densities which can not be sufficed with normal metals. High temerature superconductors are superior to low temperature superconductors due to their higher critical field, current density. High temperature superconductors (HTS) can be easily used at high temperatures and thus save a lot of cost in cryogens. Research in superconductors is fueled by its practical applocations and ability to save energy [19]. Normal metals such as copper produce heat dissipation and cause enenrgy losses due to finite resistance, which can easily be resolved through superconducting materials [20].

Even tough high temperature superconductors are superior in properties, there has been mulitple hurdles in its pracical applications. Research in superconductors has been aimed towards overcoming these difficulties. There has been postive results in small scale HTS devices that emply josephson quantum effects. Materials for HTS small scale devices are extremely perfected, but there are still issues in utilizing superconductors for integrated circuits. The field of HTS is limited by systems and the refrigeration requiremnts [47]. HTS and low temperature superconductors (LTS) small scale devices such as SQUIDS [48], passive (radio frequency) RF and microwave filters are now commercial and have applications in wide band communications and radar.

The main problem arising in HTS and LTS based devices is the refrigiration requiremnts. HTS superconductors are most researched in magnet technology for applications such as magnetic enenrgy storage, magleve trains and imaging applications such as magnetis in MRI [20]. On the basis of temperature, field and cooling requirements the application of superconductors can be divided into two categories, one category is at temperature range between 65-77K, field less than 1T and liquid nitrgen cryogen and that encompasses applications such as transformers, induction heaters, power cbales. The second category operates at 50K and 1T field and includes application such a smagleve trains, enenrgy storage devices and magnetic resonance imaging and magnetic seperaors. Some applications of superconductors in different fields are summarized in Fig.3 [15].

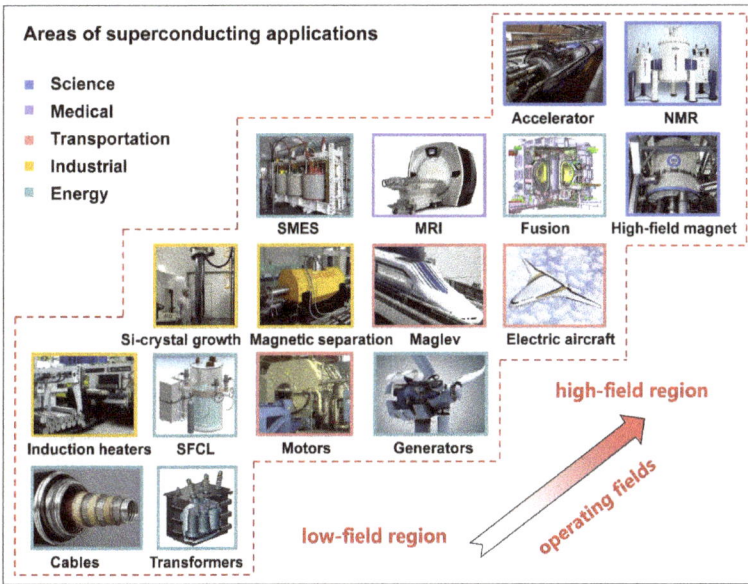

Figure 3. *Different areas of applications of superconductors with respect to operating field [15]*

2. Types of superconductors

Superconductors are a fast field which can be classified on many factors on the basis of critical temperature (high and low temperature superconductor), the coherence length, penetration depth (type I and type II superconductor), on their composition (such as oxide-based superconductors, silicon-based superconductors, carbon-based superconductors and cuperates) and on their properties such a magnetic superconductor. Only a few types are discussed briefly in this chapter, detail of these will be discussed in the coming chapters. Ginzburg and Landau posed a new theory that was inspired by Landau's general theory of 2nd order phase transitions that suggested that superconducting electrons are represented by a complex wave function ψ, that are given as

$$\frac{1}{2m}(-i\hbar\nabla + 2e\mathbf{A})^2 + (\alpha + \beta\psi\psi^*) = 0 \qquad \text{Eq.1}$$

Here **A,** is magnetic vector potential. The Ginzburg and Landau (G-L) equations give rise to two lengths, one is G-L penetration depth λ_{GL}, and other is coherence length ξ. During transition from normal to superconducting state , the superconducting current does not suddenly go from zero to maximum but instead it increases very slowly over the length

Superconductors - Materials and Applications   Materials Research Forum LLC
Materials Research Foundations 132 (2022) 49-78   https://doi.org/10.21741/9781644902110-3

called coherence length [49]. Both of these lengths depend on temperature and inversely depend on, moreover both of these quantities as temperature approached to a critical temperature, T_c. The ratio between these two lengths is expressed as Ginzburg and Landau parameter κ that is independent of temperature [49].

2.1 Type I and II superconductors

Based on Ginzburg and Landau (G-L) parameter two types of superconductors are discussed that are: Type I superconductors and Type II superconductors. If Ginzburg and Landau parameter $\kappa < \frac{1}{\sqrt{2}}$, then surface energy between metallic and conducting phase, that leads to exclusion of magnetic flu below T_c, called Meissner effect. Thus for such parameter, the superconductors are called Type I superconductors [49]. This was first noted by Abriksov in 1957.

On the other hand, $\kappa > \frac{1}{\sqrt{2}}$, in this case the surface energy is negative and flux remains within the material and thus leads to Type II superconductor. This condition is more energetically favorable. To achieve normal low energy state, the area of the boundary between normal and superconducting state is maximized, moreover area of normal material is divided to reach a quantum limit. In type II superconductors, the superconductivity doesn't breakdown at critical magnetic field, but its penetration starts at a much lower critical field, and total penetration of the field happens at an upper value of critical field. In between lower and upper critical fields , supercurrents exist that support the induvial flux quanta of the superconductor in this mixed state [50].

In type I superconductors, there exist a mixed state as the behavior changes from superconducting to normal conducing state, in the mixed state the volume of the superconducting and normal state is divided into gills. The transition from the superconducting layer which has the maximum cooper pairs and thus the maximum superconducting current to normal state which has no cooper pairs through layer whose length is determined by the coherence length. Coherence length along with GL-penetration depth can both be anisotropic in low symmetry compounds such as in low temperature superconductors.

In type II superconductors, they exist in a mixed state at an intermediate field between low and upper critical field, this state contains magnetic flux lines that are penetrating through the superconductor at this state this is considered in Shubnikov phase. The penetration of these magnetic flux lines can be identified via introduction of defects and dislocations. Th introduction of pinning defects results in the increment of critical field which leads to type III superconductors or non-ideal type II superconductors.

Type II superconductors can bear high external magnetic field and current at any temperature. Magnetization I vs H_a Curve for Type II superconductor is explained in Fig.4 [50].

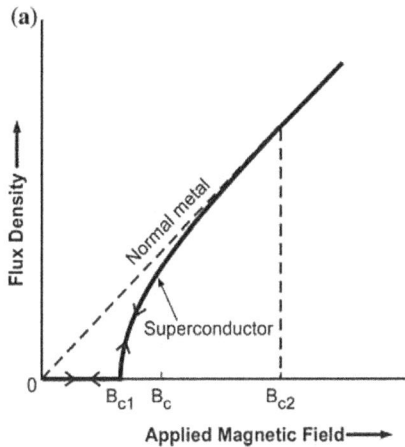

Figure 4. *Magnetization vs applied field for type II superconductor [50]*

2.2 Organic superconductors

The discovery of these superconductors was done in 1979 that contain double bonds and supply electrons for the formation of cooper pairs. Organo-metallic superconductors consist of vibrational modes to make possible phonon induced mechanism. If due to strong correlations between electrons, they become attractive, then the mechanism of phonon induced is not necessary. Most of the organic superconductors exhibit superconductivity and antiferromagnetism [51]. For organic superconductors properties depend on annealing temperature and pressure and critical temperature of such materials is mostly <10K. If stress is applied paralleled to the normal conducting planes of the organic superconductor such as k-(Bis(ethylenedithio-tetrathiafulvalene (BEDT-TTF))$_2$Cu(SCN)$_2$, this induces increment in critical temperature T_c, and H_{c2}, but a stress orthogonal to the planes to causes a decrease in these parameters. High magnetic field can destroy superconducting behavior, but that's not always the case. Superconductivity can only be seen at high magnetic field [52].

Superconductivity results when a cooper pair is formed as a consequence of an interaction between spin up and a spin down electrons. The condition for this superconductivity in organic superconductors is external magnetic field to be aligned along the layer of the organic superconductor. Crystalline λ-(BETS)$_2$ FeCl$_4$ is an organic superconductor which is antiferromagnetic at zero field below 8.5 K, but when applied field is parallel to the layers it becomes metallic at 0.8K above 41 T. A small magnetic field as 0.1 T can destroy superconductivity in this material. In organic superconductors superconductivity is strong along the c-axis as compared to cuperates superconductors. Table 2, gives some organic superconductors and some of their properties such as pressure and critical temperature [53].

Table 2. *Properties of organic superconductors with BEDT-TTF=Bis (ethylenedithio), TTF=tetrathiafulvalene [53]*

Formula	Phase	Pressure (kbar)	$T_c(K)$
$(BT)_2ReO_4$	-	3.9	2
$(BT)_2I_3$	β_L	0.0	1.5
$(BT)_2IBr_2$	β	0.0	2.3-3
$(BT)_2AuI_2$	β	0.0	3.5-5
$(BT)_2Hg_{2.89}Br_8$	-	0.0	4

2.3   Magnetic superconductors

A new class of superconductors was discovered by Chevrel and company in 1971 having large critical field and with general formula $M_xMo_6S_8$, here M represents metals or rare earth metals. Magnetic superconductors exhibit feature not exhibited by type I superconductors. When material is cooled, its superconductivity is destroyed as it transitions from superconducting to normal conducting state. When materials is cooled from normal state, it becomes superconducting below the critical temperature and if temperature is further decreased below Neel temperature, material; becomes magnetically ordered [54]. Below a specific temperature called kondo temperature, a bound state of electron is formed when a conducting electron interacts with magnetic atom. Due to this interaction, resistivity shows a minimum as a function of temperature, this is called kondo effect and it is exhibited by superconductors that contain impurities at low temperature due to destruction of superconductivity.

In magnetic superconductors, there are two lattice one is lattice of metallic atoms and the other is of magnetic atoms, the reason for the existing of two lattices is that for magnetically ordered phase to exist easily below the critical temperature [55].

2.4   High temperature superconductors (HTS)

Bednorz and Miller first discovered high temperature superconductors (HTS) in 1986, when they discovered $(La,Ba)Cu_2O_4$ with T_c of 35 K. Soon after its discovery new high temperature oxide superconductors were also observed after their discovery. The measured critical temperature for HTS is now -133.5K for the superconductor $HgBa_2Ca_2Cu_3O_{8+x}$. HTS materials have properties like type II superconductors. The behavior of HTS can be well explained with London and Ginzburg and Landau theory but no specific microscopic theory for HTS has generally been proposed and accepted. MgB_2 is a high temperature superconductor with critical temperature of 39K.

Discovery of HTS produced quite a stir in the field because of their large current above 77k. High temperature superconductors discovery opened doorways to many large scale applications along with superconducting reaching room temperature [20]. HTS materials have one or more CuO_2 planes. The physical and mechanical properties depend on crystalline and anisotropic structure, called perovskites. Normal superconductors don't have any structural effect since coherence length is larger than G-L penetration depth. But structural effects are present in HTS since coherence is very small compared to penetration depth. HTS are generally tetragonal in crystal structure. Search for high temperature superconductors, especially realization of room temperature is still very much alive [56].

3. Properties of superconductors

There are some remarkable properties of semiconducting materials which are much important for discussion and proper comprehension for current technology. The research on these properties is still ongoing to identify and make use of these properties in numerous research fields which are listed below.

- Zero Electric Resistance
- Meissner Effect
- Transition Temperature/Critical Temperature
- Josephson Currents
- Critical Current
- Persistent Currents

3.1 Zero electric resistance

Zero electrical resistance means infinite conductivity. In superconducting materials zero electrical resistance occurs at very low temperatures. When superconducting materials are cooled down below its transition temperature, the resistivity of the materials suddenly becomes zero. For example, below 4 K temperature the Mercury shows zero resistance.

For estimating electrical resistance, a new standard technique was established during the early research of superconductivity. The electrical voltage along a current-carrying sample was determined. If the resistance were absolutely zero, current will flow permanently as long as the leading ring maintained superconducting behavior when value of R is determined. However, the current will decay with time according to an exponentially decreasing law we have:

$$I(t) = I_o \, e^{-\left(\frac{R}{L}\right)\tau} \hspace{4cm} Eq.2$$

Superconductors - Materials and Applications   Materials Research Forum LLC
Materials Research Foundations 132 (2022) 49-78   https://doi.org/10.21741/9781644902110-3

At absolute zero i.e., near 273°C the conductor is cooled down to very low temperature and its resistance totally becomes zero. The most apparent property of a superconductor is the thorough loss of its electrical resistance below the certain temperature called its critical temperature. Experiments were carried out to try to sense whether a small remaining resistance exists in the superconducting state. The sensitivity test involves inducing current to flow around the superconducting ring and observing whether the current decreases. The magnetic field's magnitude is proportional to the current through the loops and it can be determined without removing the circuit's power supply.

Experiments of this kind have been carried out for many years and the magnetic field and therefore the superconducting current has continuously sustained constant in the accuracy of the measuring device. Such a continuous current is characteristic of the superconducting state. From the instance of no current reduction, it is assumed that the resistivity ρ of the superconductor materials is below the value of 10^{-26} Ω m. Resistivity means the reciprocal of the conductivity i.e ($\rho = \sigma^{-1}$). We select to define a superconductivity of superconductor by ($\rho = 0$), instead by $\sigma = \infty$.

The simplest technique to gauge the electrical resistance of some material is to put it in an electrical circuit in series with a flow source I and measure the subsequent voltage V across the material. The opposition of the material is explained with Ohm's law as R = V/I. Assuming the voltage is zero; this implies that the resistance is zero [57-59].

3.2   Meissner effect

In 1933, discovery of Meissner and Ochsenfeld showed that when temperature of superconductor is cooled to critical temperature T_c the magnetic flux is totally excluded from it, as the material does not permit to pass the magnetic field lines through it. This event in superconductors is called Meissner effect. The phenomenon is critical because it demonstrates that a bulk superconductor behaves differently under an external magnetic field 'H' as if within the specimen.

$$B = \mu_o(H + M) = 0 \text{ or } X = -1. \qquad \text{Eq.3}$$

A superconductor material, in other words, is said to be perfect diamagnetic.

3.3   Transition temperature

This is the temperature when the conducting state of superconductors changes from its normal state to superconducting state. This transition temperature is also known as critical temperature. According to the nature of the superconductors different materials have different critical temperatures. Table 3, gives critical temperature of some common superconducting materials.

Table 3. *Critical temperature values of some superconducting materials [60]*

Material	T_c [K]
(Lead)Pb	7.19
(Niobium) Nb	9.2
Nb $-$ Ti alloys	~10
Nb_3Sn	18.2
Nb_3Ge	23.3
$Y_1Ba_2Cu_3O_{7-x}$(YBCO)	90.1
$BI_2Sr_2Ca_1Cu_2O$(Bi2212)	80
$BI_2Sr_2Ca_2Cu_3O$(Bi2223)	126

3.4   Critical current

The magnetic field developed under conditions of superconductivity due to the flow of electrons through the conductor and it can be enhanced when flow of current increases beyond the value called critical value. At this critical value, the conductor comes back to its usual condition and then this value of current is called critical current. The magnetic field can be strengthened if the current flow exceeds a specific rate, which is comparable to the conductor's critical value at which it returns to its normal state.

3.5   Persistent currents

When semiconductor's ring is placed within a magnetic field at some critical temperature, then at that time temperature of the semiconductor ring will be below critical temperature. Due to self-inductance, current can be induced in the ring if this field is removed. The generated current resists the change in magnetic flux travelling over the ring, according to Lenz's law. When the ring is in a superconducting state, the flow of current remains continuous; this is referred to as the Persistent current. This persistent current will produce a magnetic flux that will keep the constant flux in the ring [61, 62].

3.6   Idealized diamagnetisms, flux lines, with its quantization

Ideal diamagnetism is a phenomenon happening in specific materials at very low temperature, specified by the complete disappearance of magnetic permeability and elimination of entire magnetic field. Super diamagnetism recognizes that phase transition is a stage of material's superconductivity. Superconducting magnetic rising is because of super diamagnetism, which resists a permanent magnet which address the superconductor, and flux restraining, which avoids the magnet fluctuating away.

Superconductors - Materials and Applications   Materials Research Forum LLC
Materials Research Foundations 132 (2022) 49-78   https://doi.org/10.21741/9781644902110-3

Fritz and Heinz London established the theory that elimination of magnetic flux is carried by screening of electrical current flowing at superconducting material surface that create the magnetic field around it which cancel out the magnetic field which is inside the superconducting material. It is known that the one of the specific characteristics of superconductors is zero resistance for current. Permanent currents established when these ideal conductors are positioned within magnetic field, and this magnetic field is present within the sample.

Because of induction, eddy currents will start to flow in the normal state when a magnetic field is applied. However, after the magnetic field has reached its maximum value and no variation occur with time, these eddy current vary as per Eq.2, and the magnetic field lines are equal inside and outside the superconductor [63-65].

The magnetic state will remain unchanged now if the perfect conductor is cooled below T_c, because only with variation of field induction currents can be generated. Now if the magnetic field is removed under T_c, then situation becomes interesting. The magnetic field is sustained in the interior of the perfect conductor. As a result, the interior of the perfect conductor has totally distinct magnetic fields based on how the ultimate condition, namely temperatures below T_c and applied magnetic field B_a, was achieved.

The behavior of a superconductor differs significantly from that of a perfect conductive material. We assume that a sample is cooled below T_c while being subjected to a magnetic field. Except for a very thin coating near the specimen surface, the magnetic field is removed completely from the interior of the superconductor only when the field is very weak. In this method, irrespective of the temporal order wherein the magnetism was provided and the material was cooled, and a perfect diamagnetic phase is obtained. Meissner and Ochsenfeld discovered perfect diamagnetism in 1933.

Magnetic flux lines are admitted in the superconductor material; each of which will transmit a magnetic flux quantum that is positioned at the ends of vertices and edges. Every flux line has a set of eddy currents for two field lines. The magnetic flux inside the flux line is generated by these currents, which are merged with the externally applied magnetic field to reduce the field between both the flux lines. As a result, flux vortices are also mentioned. The distance between the flux lines decreases as the external field B_a increases. In 1964 a team at the Atomic Research Institute Saclay used neutron diffraction to provide the initial experimental demonstration of a diffraction pattern of the magnetic flux in the Shubnikov phase. A team at the Nuclear Research Center also did stunning neutron diffraction studies on this magnetic arrangement [66, 67].

We now have a variety of technologies for detecting magnetic flux lines. The techniques frequently support one another and provide useful information concerning superconductivity. One technique is for spatially imaging magnetic objects called magneto-optics; the Faraday Effect is used in this case. A twist in the plane of polarization will be observed when linearly polarized light goes across a fine coating of a "Faraday-active" element such as a ferromagnetic garnet layer, due to a present magnetic field in it. After coating a superconducting material with Faraday active element, a transparent surface is placed on its top and then exposed with linearly polarized light. After passing through the ferromagnetic garnet layer, the light will be reflected at the superconductor and focused into a (charged coupled device) CCD camera. The magnetization from the superconductor's loops reaches the ferromagnetic garnet sheet causing the plane of polarized light to rotate. Analyzers along the CCD camera transmit only light with its polarization go ahead from the actual position. The spatial resolution of this approach is more than 1 m currently; roughly 10 photons per second may be captured, allowing for the monitoring of dynamic processes. Currently, the approach is only applicable to those superconductors that have a very cleaned and highly reflecting surfaces [68].

3.7 Flux quantization

In electron holography wave nature of electrons is utilized. A coherent electron signal is split into two waves one is carrier waveform and other is image wave, which then superimposed with each other and are identical to holography. More precisely, the magnetic flux lines encompassed by both waveforms can impact the phase difference 3connection between two waves. The magnetic flux quantization in superconductors is closely connected to the image phenomenon [69].

In 1961, Doll, Näbauer and Fairbank submitted a paper on flux quantization tests using superconducting hollowed cylinders, indicating that the magnetic field across the sphere/ring exists exclusively in variations of the flux quantization. The evolution of superconductivity was greatly influenced by these experiments. Constant currents were to be created using varying magnetic fields to investigate the potential of quantization of flux in a superconductive material, and the final magnetic flux can be calculated more efficiently than a flux quantumφ_o. Such tests are particularly difficult because of small magnitude of the flux quantum φ_o. To obtain a somewhat substantial shift in magnetic flux between stages, the flux across the ring must be kept on the scale of a few φ_o As a result, very tiny superconducting rings are needed, as the magnetic field lines required to create the constant currents would otherwise be insufficient. The final flux via the hole of the ring is "frozen-in" at start of superconductivity, thus we call such fields "freezing fields" [70, 71].

3.8 Josephson current

In an insulating medium when the superconducting materials are separated by thin film then the low resistance junction is formed to setup electron with cooper pairs. These electrons can tunnel from one junction to another and the flow of current because of this copper pair is called Josephson current.

In 1962 Brian D. Josephson noticed what might happen at a junction between two closely spaced superconductors that are isolated by a thin layer of insulating material [72]. Such sandwich structure is called "Josephson Junction". If the distance between two superconductors is reduced to about a few nanometers Cooper pairs which are responsible for superconductivity can tunnel from the insulating barrier. Due to the tunneling of Cooper pairs, Josephson current also called supercurrent could be made to flow across an insulating gap between two superconductors. This Josephson tunneling possessed not only highly fundamental quantum mechanical properties but also plays an important role in analog and digital electronics.

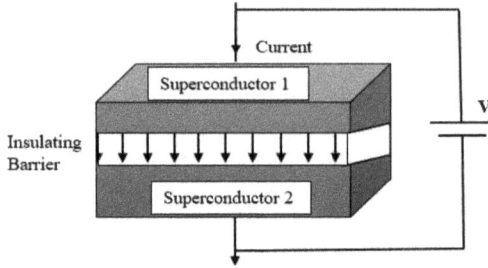

Figure 5. *The sandwich structure of two superconductors separated by thin Insulator* [73]

In his famous lecture, Feynman described the Josephson tunneling by deriving two simplified equations whose derivation is outlined here. Each superconductor of junction can be represented with its macroscopic wave function. Let Ψ_1 and Ψ_2 being two quantum-mechanical wave functions representing the superconducting state of two superconductors respectively. In the absence of a magnetic field, two wave function should be related by following time-dependent coupled Schrodinger equations.

$$i\hbar \frac{\partial \Psi_1}{\partial t} = \mu_1 \Psi_1 + K\Psi_2$$

$$i\hbar \frac{\partial \Psi_1}{\partial t} = \mu_1 \Psi_1 + K\Psi_2 \qquad \text{Eq.4}$$

K being a constant represents coupling across the barrier and μ_1, μ_2 describes the lowest energy states of each superconductor. Potential difference V can be created across the junction by connecting two superconducting regions with terminals of a battery having $\mu_1 - \mu_2 = qV$. Let's analyze these equations further by making substitutions as:

$$\Psi_1 = \sqrt{n_1} e^{i\theta_1}$$

$$\Psi_2 = \sqrt{n_2} e^{i\theta_2} \qquad \text{Eq.5}$$

Where n_1, n_2 are the density of Cooper pairs and θ_1, θ_2 are the phases on the two sides of the junction. Now by putting these substitutions in Eq.4 four equations can be obtained by equating real and imaginary parts [74].

$$\dot{n}_1 = +\frac{2}{\hbar} K\sqrt{n_2 n_1}\,\sin(\theta_2 - \theta_1) \qquad \text{Eq.7}$$

$$\dot{n}_2 = -\frac{2}{\hbar} K\sqrt{n_2 n_1}\,\sin(\theta_2 - \theta_1) \qquad \text{Eq.8}$$

$$\dot{\theta}_1 = -\frac{K}{\hbar}\sqrt{\frac{n_2}{n_1}}\cos(\theta_2 - \theta_1) - \frac{qV}{2\hbar} \qquad \text{Eq.9}$$

$$\dot{\theta}_2 = -\frac{K}{\hbar}\sqrt{\frac{n_1}{n_2}}\cos(\theta_2 - \theta_1) + \frac{qV}{2\hbar} \qquad \text{Eq.10}$$

The first two equations reveal about the kind of current that would start to flow from side1 to side2 as the time derivative of the density of Cooper pairs explains charge transport. Current that is flowing will not charge up Region 2 of the junction because currents will flow to keep the potential constant. Both n_1, n_2 will remain constant and are equal to n_0; then by setting $\frac{2Kn_0}{\hbar} = J_0$ current across the junction is just given by:

$$J = \frac{2K}{\hbar}\sqrt{n_1 n_2}\,\sin(\theta_2 - \theta_1) = J_0 \sin(\theta_2 - \theta_1) \qquad \text{Eq.11}$$

A voltage applied across the junction will shift energy levels according to $\mu_2 - \mu_1 = qV$. By putting a value of energy shifting and solving the remaining two equations of phases; what we get is:

$$\dot{\theta}_2 - \dot{\theta}_1 = \frac{qV}{\hbar} \qquad \text{Eq.12}$$

Eq.11 and Eq.12 represent important results of the theory of the Josephson junction. Here current J is called Josephson current which displays astonishing macroscopic properties in superconducting state related with phase of wave function. Because of this remarkable theoretical prediction, in 1973 Josephson got the Nobel Prize. Letting $\delta = (\theta_2 - \theta_1)$ for short; that mean Eq.12 can be written as [75]:

$$\delta(t) = \delta_0 + \frac{q}{\hbar}\int V(t)dt \qquad \text{Eq.13}$$

Here $\delta - \delta_0$ denotes gauge-invariant phase difference. To see the consequences of these important results; we will discuss some simple cases.

If DC voltages are applied across the Josephson junction voltmeter shows the voltage. As long as DC current is smaller than J_0; DC current will flow across the junction without any voltage drop. It means with zero voltages across junction current can be of any amount between $+J_0$ and $-J_0$ and will continue to flow without any resistance not only through the superconducting region but also across the barrier between them. Current will go to zero as we will try to put voltages across a junction. There is another way of getting high-frequency alternating current by applying voltages of high frequency in addition to DC voltages that exhibit strange behavior. This situation is named as the AC Josephson effect. Let

$$V = V_o + \upsilon \cos \omega t \qquad \text{Eq.14}$$

Where $\upsilon \ll V$; by using the above relation in Eq.13, we get:

$$\delta(t) = \delta_o + \frac{q}{\hbar}V_o t + \frac{q}{\hbar}\frac{\upsilon}{\omega}\sin \omega t \qquad \text{Eq.15}$$

As for small Δx

$$\sin x + \Delta x \approx \sin x + \Delta x \cos x \qquad \text{Eq.16}$$

Utilizing above approximation for $\sin \delta(t)$, final expression for Josephson current $J = J_o \sin \delta(t)$ is obtained:

$$J = J_o \left[\sin(\delta_o + \frac{q}{\hbar}V_o t) + \frac{q}{\hbar}\frac{\upsilon}{\omega}\sin \omega t \cos(\delta_o + \frac{q}{\hbar}V_o t)\right] \qquad \text{Eq.17}$$

On averaging, the first term will become zero but the second term will not only if:

$$\omega = \frac{q}{\hbar}V_o \qquad \text{Eq.18}$$

when AC voltages are applied, the Josephson current will oscillate with the above-given frequency [76] where q is a charge of Cooper pair with $q = 2q_e$. In most papers final expression for supercurrent is written in form of vector potential across junction as:

$$J = J_o \sin\left(\delta_o + \frac{2q_o}{\hbar}\int_1^2 \mathbf{A}.dl\right) \qquad \text{Eq.19}$$

Here integral upon vector potential represents path from one superconducting region to other having a small gap between them.

3.9 Josephson current in a magnetic field

When two superconductors are brought close together, there will be a unique difference between pair phases because of the exchange of electron pairs between two superconductors. By controlling the rate of pair transfer, the phase difference can be varied. With electric and magnetic fields, the coupling of pair phases between two superconducting regions can be modified. To move further one must consider single-valued wave function and phase-only changes in multiples of 2π [77] since:

$$\Psi = \sqrt{n}e^{i\theta} = \sqrt{n}e^{i(\theta+2\pi n)} \qquad \text{Eq.20}$$

Superconductors - Materials and Applications   Materials Research Forum LLC
Materials Research Foundations 132 (2022) 49-78   https://doi.org/10.21741/9781644902110-3

If the phase difference is created between two superconductors, then electron pairs will transfer in one direction resulting flow of critical current across the junction. Critical current or Josephson current also depends upon strength of magnetic flux through junction analogous with light diffraction from the double slit. Let us consider the geometry of Josephson junction in which magnetic field is passing along Z-direction along barrier height as shown in the Fig.6. Difference $\delta - \delta_o$ of two phases of Josephson junction is proportional to magnetic flux across a junction. Along X and Y direction path length is enlarged from x to x + dx and deeply inside of both superconductors for such configuration we suppose thick-walled superconductors than London penetration depth respectively.

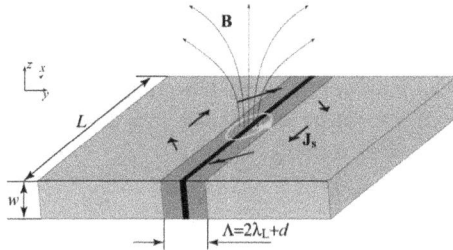

Figure 6. *Geometry of Josephson junction with effective barrier thickness with $t_{eff} = D + 2\lambda_L$ [78]*

Using the expression of quantization of fluxoid and then integrating, we get:

$$\delta(x + dx) = \mu_o \lambda_L^2 \left(\int_{1'}^{2'} \mathbf{j} d\mathbf{r} + \int_2^1 \mathbf{j} d\mathbf{r} \right) + \frac{2\pi}{\phi_o} \phi_1 \qquad \text{Eq.21}$$

where ϕ_o denotes flux quantum and ϕ_1 total flux enclosed along integrating path. λ_L is London penetration depth: beyond this depth shielding currents are exponentially small therefore integrals taken over supercurrent densities can be neglected. There is variation in Josephson current and magnetic field only along x-direction thus we are only left with [79]:

$$\frac{d\delta}{dx} = \frac{2\pi}{\phi_o} \phi_1 = \frac{2\pi}{\phi_o} B t_{eff} \qquad \text{Eq.22}$$

Where t_{eff} is an effective thickness that is obtained by integrating along specific axis. Variation in magnetic field is observed exponentially across junction within penetration depth λ_L yielding:

$$t_{eff} = 2\lambda_L + D \qquad \text{Eq.23}$$

Where London penetration depths for both superconductors are denoted by $2\lambda_L$. D is the thickness of barrier that is smaller than λ_L. In Eq.22 magnetic flux is:

$$\phi_1 = Bt_{eff}dx \qquad \text{Eq.24}$$

Furthermore, self-field generated by Josephson junction can be neglected representing lengths of sides of junction cannot be increased from Josephson penetration depth.

$$\lambda_J = \sqrt{\frac{\phi_o}{2\pi\mu_o j_c l_{eff}}} \qquad \text{Eq.25}$$

Here μ_o and j_c represent permeability and critical supercurrent density respectively. $l_{eff} = t_{eff}$ only when we have supposed superconducting's electrodes are thicker than λ_L. Josephson penetration depth can be increased up to some scale only for small quantity of supercurrent density. Integrating Eq.22 we will get:

$$\delta(x) = \delta(0) + \frac{2\pi}{\phi_o} Bt_{eff}x \qquad \text{Eq.26}$$

We assume a and b represent the length and width of the junction respectively. Maximum quantity of flowing Josephson current across junction can be calculated; firstly, by putting function $\delta(x)$ into first Josephson Eq.11 then integrating over an area of junction as:

$$I = \int_0^b dy \int_0^a dx \, J_o \sin\left[\delta(0) + \frac{2\pi}{\phi_o} Bt_{eff}x\right] \qquad \text{Eq.27}$$

Now assuming critical supercurrent density spatially homogeneous then integrating and inserting integration limits finally we will obtain:

$$I = J_o ab \sin\delta' \; \frac{\sin\left[\frac{\pi}{\phi_o} Bt_{eff}\, a\right]}{\frac{\pi}{\phi_o} Bt_{eff}\, a} \qquad \text{Eq.28}$$

Where variable $\delta' = \delta(0) + \frac{\pi}{\phi_o} Bt_{eff}a$ will adjust itself in such a way that Eq.28 is satisfied that is possible only when $\delta' = \pm 1$. Finally, we obtain a critical current of Josephson junction [2] as:

$$I(B) = I_c(0) \left| \frac{\sin\left(\frac{\pi\phi_j}{\phi_o}\right)}{\frac{\pi\phi_j}{\phi_o}} \right| \qquad \text{Eq.29}$$

Where $\phi_j = Bt_{eff}\, a$ representing magnetic flux penetrating across Josephson junction and $I_c(0) = J_o ab$. According to Eq.26 magnetic field can change gauge-invariant phase difference along barrier while supercurrent will then oscillate linearly along x-coordinates

[80]. If the external magnetic field is zero then in both directions of barrier an equal amount of supercurrent as shown by the length of colored arrows will starts to flow. At zero value of magnetic field wavelength of oscillations of supercurrent is an integer multiple of width of junction and integral over supercurrent is zero. When an applied field has any value then the wavelength of supercurrent density will not fit nearly with a width of the junction; while the supercurrent with some phase shift $\delta(0)$ having a finite value will adjust up to a certain maximum value.

When magnetic field is increased up to $\frac{\Phi_o}{2}$ then the wavelength of Josephson junction becomes equal to twice the length of junction such that only half of the wavelength fits exactly into the length of the junction. For magnetic field $\boldsymbol{\phi_o}$ full wavelength fits exactly into the length of junction [81]. But now current will reverse its direction once resulting flow of equal amount of current in both directions; thus, net current across the junction will be zero. So, the maximum value for which supercurrent is adjustable with a width of the junction becomes smaller over an increasing number of periods.

3.10 Superconducting quantum interference device (SQUID)

SQUID by using the Josephson effect and flux quantization phenomena measures small changes in magnetic flux. It is a convertor of magnetic flux into voltages and can be used to sense any quantity such as voltage, current, position, etc. that then can be transduced into a magnetic flux [82]. SQUID simply contains a closed superconducting loop having two Josephson junctions are connected in parallel on it as a and b junctions. At ends R and S any electrical instruments are connected to measure the flow of current. As junctions are identical the total current J_{total} will split equally half on each side with phases δ_a and δ_b as:

$$J_{total} = J_1 + J_2 \qquad \text{Eq.30}$$

Whether we go to any route phase difference of wave functions between R and S must be the same. Such as along upper and lower routes "a and b" phase differences between R and S are δ_a and δ_b plus line integral of the vector potential and can be written as:

$$\text{Phase change} = \delta_a + \frac{2q_o}{\hbar} \int \mathbf{A} \cdot dl$$

$$\text{Phase change} = \delta_b + \frac{2q_o}{\hbar} \int \mathbf{A} \cdot dl \qquad \text{Eq.31}$$

We have written the above two equations by integrating equation $\hbar\boldsymbol{\nabla}\theta = q\mathbf{A}$ along some path. Eq.31 must be equal; then subtracting them difference of deltas will be equal to:

$$\delta_b - \delta_a = \frac{2q_o}{\hbar} \oint \mathbf{A} \cdot dl \qquad \text{Eq.32}$$

Here integral is around all closed loop that passes through both junctions. Difference of deltas will become equal to $2q_o/\hbar$ times magnetic flux $\boldsymbol{\phi}$ as integral over vector potential provides magnetic flux across loop which passes through both branches of the junction.

$$\delta_b - \delta_a = \frac{2q_o}{\hbar}\phi \qquad \text{Eq.33}$$

The phase difference can be adjusted by changing the magnetic field whose adjusted value will help us in knowing whether or not the total current shows any interference across two junctions. For convenience we will write:

$$\delta_a = \delta_o + \frac{q_o}{\hbar}\phi, \qquad \delta_b = \delta_o - \frac{q_o}{\hbar}\phi \qquad \text{Eq.34}$$

Now using the first Josephson equation and then Eq.34 into Eq.30 we will get:

$$\begin{aligned} J_{total} &= J_o\left\{\sin\left(\delta_o + \frac{q_o}{\hbar}\phi\right) + \sin\left(\delta_o + \frac{q_o}{\hbar}\phi\right)\right\} \\ &= 2J_o\sin\delta_o\cos\frac{q_o\phi}{\hbar} \end{aligned} \qquad \text{Eq.35}$$

δ_o will depend upon externally applied voltages and $\sin\delta_o$ should not be greater than 1. Now the maximum current that can flow through SQUID is given by [74]:

$$J_{max} = 2J_o\left|\cos\frac{q_o}{\hbar}\phi\right| \qquad \text{Eq.36}$$

As a function of an applied magnetic field the maximum supercurrent within SQUID oscillates periodically. It is analogous with the light diffraction by a double slit. A SQUID can measure changes in the magnetic field down to $10^{-6}\phi$ [2]. Interference between junctions can be utilized to make a sensitive magnetometer that measures applied magnetic field sensitively. If we set a greater number of junctions that are equally spaced then for change in a magnetic field, we will get very sharp maxima and minima making more precise SQUIDs [75].

3.11 Superconductivity: A macroscopic quantum phenomenon

We are accustomed to appearing the distinct states in atoms. Quantum mechanically stationary atomic states are characterized by the angular momentum that is in integral multiples of $\hbar = h/2\pi$. Quantum mechanical wave function represents the probability of finding the electron, should be single-valued causing the angular momentum to be quantized. During our movement starting from a given place around the atomic nucleus, the wave function starts to duplicate itself when we go back to that place. Because of no effect on the wave function, the phase of the wave function can also change by an integer multiple of 2π.

On a macroscopic scale, we might have the same problem. For example, we assume an arbitrary wave travelling in a circle of radius R with no damping. If n integer multiples of wavelengths perfectly fit within the circle, the wave can become stationary. Then, using the wavenumber $k = 2\pi/\lambda$, we get the relation $n\lambda = 2\pi R$ or $kR = n$. If this condition is disrupted, the wave will vanish after a few revolutions because of interference.

By keeping this example in mind we assume a propagating electron wave around the circle. Firstly, we have to use Schrödinger equation for the appropriate configuration to get a precise treatment. We, on the other hand, refrain from doing so and instead constraint ourselves to a semi-classical approach that yields the same core consequences.

We begin by looking at the relationship between the electron's momentum and its wave vector. According to de Broglie, we get $\mathbf{p}_{kin} = \hbar\mathbf{k}$ for an uncharged quantum particle, where $\mathbf{p}_{kin} = m\mathbf{v}$ signifies the "kinetic momentum" (where v is the particle's velocity and m is the mass). This gives us the particle's kinetic energy: $E_{kin} = (\mathbf{p}_{kin})^2/2m$. According to quantum physics, the wave vector $\mathbf{k}$ of a charged carring particle like the electron is determined by the quantity vector potential $\mathbf{A}$. This vector potential is associated with the magnetic field by following relation.

$$\text{curl } \mathbf{A} = \mathbf{B} \qquad\qquad \text{Eq.37}$$

Now canonical momentum is defined as;

$$\mathbf{p}_{can} = m\mathbf{v} + q\mathbf{A} \qquad\qquad \text{Eq.38}$$

In the above relation, m and q represents mass and charge of particle respectively. Now $\mathbf{p}_{can}$ and wave vector $\mathbf{k}$ are related as;

$$\mathbf{p}_{can} = \hbar\mathbf{k} \qquad\qquad \text{Eq.39}$$

We now need that the ring has an integer number of wavelengths. Taking integral equal to an integer multiple of 2π and then integrating k along an integrating path around the circle we will get:

$$n\cdot 2\pi = \oint \mathbf{k}d\mathbf{r} = 1/\hbar \oint \mathbf{p}_{can} \, d\mathbf{r} = m/\hbar \oint \mathbf{v}d\mathbf{r} + q/\hbar \oint \mathbf{A}d\mathbf{r} \qquad \text{Eq.40}$$

The integral over area F encompassed by circle $\int_F \text{curl}\mathbf{A}d\mathbf{f}$ can replace the second integral (Adr) on the other side, according to Stokes' theorem. This replaced integral, however, only denotes magnetic flux surrounded by the circle, $\int_F \text{curl}\mathbf{A}d\mathbf{f} = \int_F \mathbf{B}d\mathbf{f} = \Phi$. As a result, Eq.40 may be transformed into

$$nh/q = m/q \oint \mathbf{v}d\mathbf{r} + \Phi \qquad\qquad \text{Eq.41}$$

Multiplying $\hbar/q$ and utilizing $h/2\pi$ in Eq.40 we get Eq.41.

We utilized a quantum condition that links the magnetic flux of circle with Planck's constant and the particle's charge in this way. The magnetic flux across the circle varies precisely by a multiple of h/q if the integral is constant on the right-hand side of Eq.41.

So far, we've only talked about one particle. What happens, though, if all or a large number of charge carriers are in the same quantum state? There is also need to represent charge carriers as a single coherent wave having a well-defined phase, in which all charge carriers change their quantum states at the same time. In this scenario, Eq.41 applies to this coherent matter wave as well.

3.12   Critical magnetic field

When the magnetic field (either external or created by the current flowing superconductor itself) exceeds a particular value, the superconducting state / phase of a superconducting material breaks and the sample begins to behave like an ordinary conductor. The critical magnetic field represents magnitude of magnetic field beyond which a superconductor recovers to its normal conducting state. The critical magnetic field value varies with temperature. The value of the critical magnetic field increases as the temperature (below the critical temperature) decreases [83].

Conclusion

In this chapter, we have discussed a novel class of materials called superconductors that show outstanding properties below a certain temperature such as zero resistance, ideal diamagnetism, and perfect conductivity. These properties along with many other inspiring factors make superconductivity and its field attractive for research. The chapter discussed, superconductors with its two types denoted by type I and type II and high T_c superconductors briefly. Properties of superconductors such as zero resistance, procedures to detect zero resistance, critical temperature, Current, internal behavior, expulsion of magnetic field also called Meissner effect, how flux lines are expelled by eddy currents, detection of flux lines are deliberated. A unique approach to detect flux lines that employs magneto-optic effect and a ferromagnetic garnet layer over the superconducting layer is discussed. This technique is valid for only very high reflective superconducting materials. Flux quantization is a phenomenon that happens in superconductors that resembles electron holography.

Josephson junction formed by two superconducting thin films with an insulating barrier between them, a resistance-less Josephson current can pass through a very thin insulating barrier between two superconducting regions making it useful for a variety of applications. When a junction is made two effects are studied first is that for DC flow of current; current will continue to flow even voltages dropped across junction are zero. Second, for AC flow of current Josephson current will oscillate with some frequency. When an externally applied magnetic field is increased across junction supercurrent in junction varies periodically and may reverse its direction at several points. The most widely used application of the Josephson effect is SQUID which is explained here for two Josephson junctions. The maximum current that flows within SQUID helped us in measuring changes in magnetic field sensitively has also been calculated. However, if we set equally spaced multi junctions that are then for change in a magnetic field, we will get very sharp maxima and minima making more precise SQUIDs.

References

[1] B. Holder, A. H. Keller, High-temperature superconductors: underlying physics and applications, Z. Naturforsch. B. 75(2020) 3-14. https://doi.org/10.1515/znb-2019-0103

[2] W. Buckel, R. Kleiner, Fundamental properties of superconductors in Superconductivity: Fundamentals and Applications, second ed., JWS., Germnay, 2008 , pp. 11-71.

[3] B. Seeber, Handbook of Applied Superconductivity, first ed., CRC press, Boca, Rtaon , 1998, p. 175-77 https://doi.org/10.1887/0750303778

[4] J. Bardeen, L.N. Cooper, J.R. Schrieffer, Theory of superconductivity, Phys. Rev. 8(1957) 1170- 1175. https://doi.org/10.1103/PhysRev.108.1175

[5] L.N. Cooper, and D. Feldman, BCS: 50 years, Mod. Phys. Lett., Vol 25, 2010, pp. 3169-3189 https://doi.org/10.1142/S0217732310034626

[6] J.G. Bednorz, K.A. Müller, Possible high T c superconductivity in the Ba− La− Cu− O system, Zeitschrift für Physik B Condensed Matter. 64(1986) 189-193. https://doi.org/10.1007/BF01303701

[7] W. Buckel, R. Kleiner, Superconductivity: Fundamentals and Applications, second ed., JWS., Germnay, 2008 , pp. 11-71.

[8] G. Revathy, V. Rajendran, P.S. Kumar, Prediction study on critical temperature (c) of different atomic numbers superconductors (both gaseous/solid elements) using machine learning techniques, Mater. Today: Proc. 44(2021) 3627-3632. https://doi.org/10.1016/j.matpr.2020.10.091

[9] Y. Maeno, H. Hashimoto, K. Yoshida, S. Nishizaki, T. Fujita, J.G. Bednorz, F. Lichtenberg, Superconductivity in a layered perovskite without copper, Nature. 72(1994) 532-534. https://doi.org/10.1038/372532a0

[10] N. Nagaosa, Superconductivity and antiferromagnetism in high-TC cuprates, Science. 275(1997) 1078-1079. https://doi.org/10.1126/science.275.5303.1078

[11] K. Prassides, Y. Iwasa, T. Ito, D.H. Chi, K. Uehara, E. Nishibori, M. Takata, M. Sakata, Y. Ohishi, O. Shimomura, T. Muranaka, Compressibility of the MgB 2 superconductor, Phys. Rev. B. 64(2001) 012508- 012509. https://doi.org/10.1103/PhysRevB.64.012509

[12] Y. Kamihara, H. Hiramatsu, M. Hirano, R. Kawamura, H. Yanagi, T. Kamiya, H. Hosono, Iron-based layered superconductor: LaOFeP, J. Amer. Chem.Soc. 128(2006) 10012-10013. https://doi.org/10.1021/ja063355c

[13] A. Drozdov, M.I. Eremets, I.A. Troyan, V. Ksenofontov, S.I. Shylin, Conventional superconductivity at 203 kelvin at high pressures in the sulfur hydride system, Nature. 525(2015) 73-76. https://doi.org/10.1038/nature14964

[14] E. Snider, N. Dasenbrock-Gammon, R. McBride, M. Debessai, H. Vindana, K. Vencatasamy, K.V. Lawler, A. Salamat, R.P. Dias, Room-temperature superconductivity in a carbonaceous sulfur hydride, Nature. 586(2020) 373-377. https://doi.org/10.1038/s41586-020-2801-z

[15] C. Yao, Y. Ma, Superconducting materials: Challenges and opportunities for large-scale applications, Iscience. (2021) 102541. https://doi.org/10.1016/j.isci.2021.102541

[16] H. Rogalla, and P.H. Kes, 100 years of superconductivity, T&F. (2011) https://doi.org/10.1201/b11312

[17] L. Boeri, R.G. Hennig, P.J. Hirschfeld, G. Profeta, A. Sanna, E. Zurek, W. E. Pickett, M. Amsler, R. Dias, M. Eremets, C. Heil, The 2021 room-temperature superconductivity roadmap, J. Condens. Matter Phys. (2021)

[18] A. Bianconi, N. Poccia, Superstripes and complexity in high-temperature superconductors, J. Supercond. Nov. Magn. 25(2012) 1403-1412. https://doi.org/10.1007/s10948-012-1670-6

[19] M. Chen, L. Donzel, M. Lakner, W. and Paul, High temperature superconductors for power applications, J. Eur. Ceram. Soc. 24(2004) 1815-1822. https://doi.org/10.1016/S0955-2219(03)00443-6

[20] T. Silver, S. Dou, J. Jin, Applications of high temperature superconductors, Europhys. News. 32(2001) 82-86. https://doi.org/10.1051/epn:2001302

[21] J.F. Cochran, D. Mapother, Superconducting transition in aluminum, Phys.Rev. 111(1958) 130- 132. https://doi.org/10.1103/PhysRev.111.132

[22] I. Belash, O. Zharikov, A. Palnichenko, Superconductivity of GIC with Li, Na and K, Synth. Met. 34(1989) 455-460. https://doi.org/10.1016/0379-6779(89)90424-4

[23] B. Matthias, TH Geballe und VB Compton, Rev. Mod. Phys. 35(1963) 413-414 https://doi.org/10.1103/RevModPhys.35.414.2

[24] N. Emery, C. Herold, J.F. Marêché, P. Lagrange , Synthesis and superconducting properties of CaC6, Sci. Technol. Adv. Mater. (2009) 3-6

[25] 竹内大輔 山崎聡, ダイヤモンド表面の全光電子放出率分光法 (TPYS). 表面科学, 29(2008) 151-158.

[26] K. Tanigaki, T.W. Ebbesen, S. Saito, J. Mizuki, J.S. Tsai, Y. Kubo ,S. Kuroshima , Superconductivity at 33 K in Cs x Rb y C 60, Nature. 352(1991) p. 222-223. https://doi.org/10.1038/352222a0

[27] M. J. Rosseinsky, AP. Ramirez ,S.H. Glarum,D.W. Murphy, R.C. Haddon, A.F. Hebard, T.T.M.Palstra, A.R.Kortan, S.M. Zahurak, A.V.Makhija, Superconductivity at 28 K in RbxC60. Phys. Rev. Lett. 66(1991), 2825-2830.

[28] T. Inushima, Electronic structure of superconducting InN, Sci. Technol. Adv. Mater. 7(2006)S112-S116. https://doi.org/10.1016/j.stam.2006.06.004

[29] K. Makise, N. Kokubo, S. Takada, T. Yamaguti, S. Ogura, K. Yamada, B. Shinozaki, K. Yano, K. Inoue, H. and Nakamura, Superconductivity in transparent zinc-doped In2O3 films having low carrier density, Sci. Technol. Adv. Mater.9(2009) 044202- 044208. https://doi.org/10.1088/1468-6996/9/4/044208

[30] G. Schell, H. Winter, H. Rietschel, F. Gompf, Electronic structure and superconductivity in metal hexaborides, Phys.Rev. B. 25(1982)1586-1589. https://doi.org/10.1103/PhysRevB.25.1589

[31] J. Nagamatsu, et al., Superconductivity at 39 K in magnesium diboride, Nature. 410(2001) 63-64. https://doi.org/10.1038/35065039

[32] K. - H. Bernhardt, Preparation and superconducting properties of niobium carbonitride wires, Z. Naturforsch. 30(1975) 528-532. https://doi.org/10.1515/zna-1975-0422

[33] G.-i. Oya, and E. Saur, Preparation of Nb3Ge films by chemical transport reaction and their critical properties, J. Low Temp. Phys. 34(1979) 569-583. https://doi.org/10.1007/BF00114941

[34] R. D. Fowler, B.T. Matthias, L.B. Asprey, H.H. Hill, J.D.G. Lindsay, C.E. Olsen, R.W. White, Superconductivity of protactinium, Phys.Rev. Lett. 15(1965) 860. https://doi.org/10.1103/PhysRevLett.15.860

[35] J. Hulm, C.K. Jones, R.A. Hein, J.W. and Gibson, Superconductivity in the TiO and NbO systems, J. Low Temp. Phys. 7(1972) 291-307. https://doi.org/10.1007/BF00660068

[36] J. Eisenstein, Superconducting elements, Rev. Mod. Phys. 26(1954) 273- 277. https://doi.org/10.1103/RevModPhys.26.277

[37] B. Matthias, and T. Geballe, s. Geller, and E. Corenzwit, Superconductivity of nb 3 sn, Phys. Rev, 95(1954) 1430-1435. https://doi.org/10.1103/PhysRev.95.1435

[38] T. Muranaka, Y. Kikuchi, T. Yoshizawa, N. Shirakawa, J. and Akimitsu, Superconductivity in carrier-doped silicon carbide, Sci. Technol. Adv. Mater., 9(2009) 044200- 044204. https://doi.org/10.1088/1468-6996/9/4/044204

[39] H. O. Pierson, Handbook of refractory carbides and nitrides: properties, characteristics, processing and applications, first ed., William Andrew., New York, 1996, pp. 350-362 https://doi.org/10.1016/B978-081551392-6.50017-9

[40] S. Tanaka, A. Miyake, T. Kagayama, K. Shimizu, P. Burger, F. Hardy, C. Meingast, Y. Ōnuki, Superconducting and Martensitic Transitions of V3Si and Nb3Sn under High Pressure, J. Phys. Soc. Jpn., 81(2012) SB024-SB026. https://doi.org/10.1143/JPSJS.81SB.SB026

[41] Z. Fisk, P. Schmidt, and L. Longinotti, Growth of YB6 single crystals, MRS Bulletin. 11(1976) 1019-1022. https://doi.org/10.1016/0025-5408(76)90179-3

[42] W. Lengauer, Characterization of nitrogen distribution profiles in fcc transition metal nitrides by means of Tc measurements, Surf. Interface Anal.15(1990) 377-382. https://doi.org/10.1002/sia.740150606

[43] M.I. Tsindlekht, V.M. Genkin, G.I. Leviev, I. Felner, O. Yuli, I. Asulin, O. Millo, M.A. Belogolovskii, N.Y. and Shitsevalova, Linear and nonlinear low-frequency electrodynamics of surface superconducting states in an yttrium hexaboride single crystal, Phys. Rev. B. 78(2008) 024520- 024522. https://doi.org/10.1103/PhysRevB.78.024522

[44] L. Malavasi, U.A. Tamburini, P. Galinetto, P. Ghigna, G. Flor, The High-Temperature Superconductor EuBa2Cu3O+x: Role of Thermal History on

Microstructure and Superconducting Properties, J. mater. synth.Process., 9(2001) 31-37. https://doi.org/10.1023/A:1011334631235

[45] Y. Shi, N.H. Babu, K. Iida, D.A. Cardwell, Superconducting properties of Gd-Ba-Cu-O single grains processed from a new, Ba-rich precursor compound. in J. Phys. Conf. Ser., IOP Publishing. (2008) https://doi.org/10.1088/1742-6596/97/1/012250

[46] A. E. Lita, D. Rosenberg, S. Nam, A.J. Miller, D. Balzar, L.M. Kaatz,R.E. Schwall, Tuning of tungsten thin film superconducting transition temperature for fabrication of photon number resolving detectors, IEEE transactions on applied superconductivity, 15(2005) 3528-3531. https://doi.org/10.1109/TASC.2005.849033

[47] M. Baecker, Energy and superconductors-applications of high-temperature-superconductors, Oldenbourg Wissenschaftsverlag. (2011) 343-351

[48] J. Clarke, and A.I. Braginski, The SQUID handbook: Applications of SQUIDs and SQUID systems, JWS., Germany, 2006. pp. 29-92 https://doi.org/10.1002/9783527609956

[49] P. Schmüser, Superconductivity in high energy particle accelerators, Prog. Part. Nucl. Phys. 49(2002) 155-244. https://doi.org/10.1016/S0146-6410(02)00145-X

[50] R. G. Sharma, Superconductivity: Basics and applications to magnets, first ed., Springer Nature., Vol 214, 2021, pp. 620-631 https://doi.org/10.1007/978-3-030-75672-7

[51] A. Gabovich, and A. Voitenko, Superconductors with charge-and spin-density waves: theory and experiment, Low Temp.Phys. 26(2000) 305-330. https://doi.org/10.1063/1.593902

[52] J. Singleton, Studies of quasi-two-dimensional organic conductors based on BEDT-TTF using high magnetic fields, Rep. Prog. Phys. 63(2000), 1109-1111. https://doi.org/10.1088/0034-4885/63/8/201

[53] L. A. Parinov, Microstructure and properties of high-temperature superconductors, Springer Science & Business Media. (2013) https://doi.org/10.1007/978-3-642-34441-1

[54] A. Mourachkine, High-temperature superconductivity in cuprates: the nonlinear mechanism and tunneling measurements, Springer Science & Business Media, 125(2002) https://doi.org/10.1007/0-306-48063-8

[55] D. C. Johnston, and H.F. Braun, Systematics of superconductivity in ternary compounds, in Superconductivity in ternary compounds II, Springer. (1982) 11-55. https://doi.org/10.1007/978-1-4899-3768-1_2

[56] H. Kotegawa, Y. Tokunaga, K. Ishida, G.Q. Zheng, Y. Kitaoka, H. Kito, A. Iyo, K. Tokiwa, T. Watanabe, H. Ihara, Unusual magnetic and superconducting characteristics in multilayered high-T c cuprates: 63 Cu NMR study, Phys.Rev. B. 64(2001) 064510-064515. https://doi.org/10.1103/PhysRevB.64.064515

[57] D. Quinn III, and W. Ittner III, Resistance in a Superconductor, J.App. Phys. 33(1962) 748-749. https://doi.org/10.1063/1.1702504

Superconductors - Materials and Applications
Materials Research Foundations 132 (2022) 49-78

Materials Research Forum LLC
https://doi.org/10.21741/9781644902110-3

[58] Tuyn, W. and H.K. Onnes, Further experiments with liquid helium, AA. The disturbance of supra-conductivity by magnetic fields and currents. The hypothesis of Silsbee, in Through Measurement to Knowledge, Springer, (1991)363-387. https://doi.org/10.1007/978-94-009-2079-8_23

[59] C. J. Gorter, and H. Casimir, On supraconductivity , Physica. 1(1934) 306-320. https://doi.org/10.1016/S0031-8914(34)90037-9

[60] P. Lebrun, L. Tavian, U. Wagner, G. Vandoni, Cryogenics for particle accelerators and detectors, (2002)

[61] B. Serin, Superconductivity. Experimental part, in Low Temperature Physics II/Kältephysik II, Springer. (1956) 210-273. https://doi.org/10.1007/978-3-642-45838-5_3

[62] P. F. Dahl, Superconductivity after World War I and circumstances surrounding the discovery of a state B= 0, Historical Studies in the Physical and Biological Sciences, 16(1986) 1-58. https://doi.org/10.2307/27757556

[63] P. E. Goa, H. Hauglin, M. Baziljevich, E. Il'yashenko, P.L. Gammel, T.H. Johansen, Real-time magneto-optical imaging of vortices in superconducting NbSe2, Supercond Sci Technol.14(2001) 720- 729. https://doi.org/10.1088/0953-2048/14/9/320

[64] T. Matsuda, S. Hasegawa, M. Igarashi, T. Kobayashi, M. Naito, H. Kajiyama,J. Endo, N. Osakabe, A. Tonomura, R. Aoki, 1Magnetic field observation of a single flux quantum by electron-holographic interferometry, Phys. Rev. Lett. 62(1989) 2512-2519. https://doi.org/10.1103/PhysRevLett.62.2519

[65] R. Straub, S. Keil, R. Kleiner, D. Koelle, Low-frequency flux noise and visualization of vortices in a YBa2Cu3O7 dc superconducting quantum interference device washer with an integrated input coil, App. Phys.Lett. 78(2001) 3645-3647. https://doi.org/10.1063/1.1378048

[66] D. Cribier, et al., Mise en evidence par diffraction de neutrons d'une structure periodique du champ magnetique dans le niobium supraconducteur, Phys. Lett. 9(1964)106-107. https://doi.org/10.1016/0031-9163(64)90096-4

[67] V. Gantmakher, Progress in Low Temperature Physics, (1967)

[68] J. Schelten, H. Ullmaier, and W. Schmatz, neutron diffraction by vortex lattices in superconducting nb and Nb0.73Ta0.2, Kernforschungsanlage, Juelich. Ger. (1971) https://doi.org/10.1002/pssb.2220480219

[69] M. McCartney, R. Dunin-Borkowski, and D. Smith, Electron holography of magnetic nanostructures, in Magnetic microscopy of nanostructures, Springer. (2005) 87-109. https://doi.org/10.1007/3-540-26641-0_5

[70] R. Doll, and M. Näbauer, Experimental proof of magnetic flux quantization in a superconducting ring, Phys. Rev. Lett.7(1961) 49- 51. https://doi.org/10.1103/PhysRevLett.7.51

[71] Jr. Deaver, B.S. and W.M. Fairbank, Experimental evidence for quantized flux in superconducting cylinders, Phys. Rev. Lett. 7(1961) 40- 43. https://doi.org/10.1103/PhysRevLett.7.43

[72] B. D. Josephson, Possible new effects in superconductive tunnelling, Phys. Lett.1(1962) 251-253. https://doi.org/10.1016/0031-9163(62)91369-0

[73] R. Kleiner, and W. Buckel, Superconductivity: an introduction, JWS. (2016) https://doi.org/10.1002/9783527686513

[74] A. Kraft, C. Rupprecht, and Y.-C. Yam, Superconducting Quantum Interference Device (SQUID). 2017, University of British Columbia.

[75] R. P. Feynman, R.B. Leighton, and M. Sands, The feynman lectures on physics; vol. i, Amer.J. Phys. 33(1965) 750-752. https://doi.org/10.1119/1.1972241

[76] S. Shapiro, Josephson currents in superconducting tunneling: The effect of microwaves and other observations, Phys. Rev.Lett. 11(1963) 75- 80. https://doi.org/10.1103/PhysRevLett.11.80

[77] L. N. Cooper, Microscopic quantum interference in the theory of superconductivity, Science. 181(1973) 908-916. https://doi.org/10.1126/science.181.4103.908

[78] Jr, A. Abdumalikov et al., Nonlocal electrodynamics of long ultranarrow Josephson junctions: experiment and theory, Phys. Rev. B.74(2006) 134510- 134515. https://doi.org/10.1103/PhysRevB.74.134515

[79] J. F. I. Nturambirwe, Superconducting quantum interference device (SQUID) magnetometers: Principles, fabrication and applications, Postgraduate diploma assay, (2010)

[80] N. Byers, and C. Yang, Theoretical considerations concerning quantized magnetic flux in superconducting cylinders, Phys. Rev. Lett.7(1961) 40- 46. https://doi.org/10.1103/PhysRevLett.7.46

[81] D. N. Langenberg, D.J. Scalapino, and B.N. Taylor, The Josephson Effects., Sci.Amer. 214(1966) 30-39. https://doi.org/10.1038/scientificamerican0566-30

[82] R. Fagaly, Superconducting quantum interference device instruments and applications, Review of scientific instruments, 77(2006) 1011094-101101. https://doi.org/10.1063/1.2354545

[83] V. L. Ginzburg, and L.D. Landau, On the theory of superconductivity, in On superconductivity and superfluidity, Springer. (2009)113-137. https://doi.org/10.1007/978-3-540-68008-6_4

Superconductors - Materials and Applications

Materials Research Foundations 132 (2022) 79-96

Materials Research Forum LLC

https://doi.org/10.21741/9781644902110-4

Chapter 4

Superconductors for Large-Scale Applications

Subhojit Bose[1], Sarit Chakraborty[1], Biplab Roy[2] and Pinku Chandra Nath[3*]

[1]Department of Physics, National Institute of Technology Agartala, Jirania, Tripura, India-799046

[2]Department of Chemical Engineering, National Institute of Technology Agartala, Jirania, Tripura, India-799046

[3]Department of Bio Engineering, National Institute of Technology Agartala, Jirania, Tripura, India-799046

* nathpinku005@gmail.com

Abstract

The discovery of superconducting material and its synthesis leads to several advantages application in the modern developing society. The advantages of superconducting material cover a broad area of research. In superconductors, one of the characteristics that distinguish them is their critical temperature. This temperature specifies the range and kind of superconducting materials that may be used in industrial applications and generate profit. In this chapter, some of the advantageous applications of semiconducting materials are discussed that includes the magnetic levitation train, the nuclear magnetic resonance, superconducting magnetic energy storage (SMES), and the transmission of electrical energy. Furthermore, advantageous use of fast-field cyclic nuclear magnetic resonance which benefits from superconducting blocks is presented. The unexpected discovery of superconducting materials also sparked the research in fundamental physics, material science and advantageous technological application to improve the quality of daily human life.

Keywords

Superconductors, SMES, Electrical Energy, Semiconducting Materials, Transmission

Abbreviations

LTS, Low Temperature Superconductor.
NMR, Nuclear Magnetic Resonance.
LHC, Large Hadron Collider.
MRI, Magnetic Resonance Imaging
SMES, Superconducting Magnetic Energy Storage.
FES, Flywheel Energy Storage.
HTS, High Temperature Superconductor.

Superconductors - Materials and Applications   Materials Research Forum LLC
Materials Research Foundations 132 (2022) 79-96   https://doi.org/10.21741/9781644902110-4

SMB, Superconducting magnetic bearings.
EDS, Electrodynamics Suspension System.
SCMaglev, Superconducting Maglev.
FFC, Fast Field Cycling.
SQUID, Superconducting Quantum Interference Device.

Contents

1.   Introduction

Heike Kamerling Onnes proposed the concept of superconductivity in the year of 1911, a quantum phenomenon characterise by zero electrical resistance [1]. The development of superconductivity was completely unintentional, and the primary goal of the experiment was to liquefy the helium noble gas [2]. Superconducting is now emerging as an intriguing technology with real world applications. For a superconductor the resistance almost becomes zero above the absolute zero temperature and this is known as critical temperature (T_c) [3, 4]. It was observed that in application of magnetic field, the critical temperature

Superconductors - Materials and Applications   Materials Research Forum LLC
Materials Research Foundations 132 (2022) 79-96   https://doi.org/10.21741/9781644902110-4

decreases and with sufficient strength of magnetic field the superconductivity vanishes and material behaves like a normal conductor [5]. It is possible to degrade the superconducting phase by applying a strong magnetic field with an intensity that is less than the threshold level of magnetic field intensity and the material undergoes a phase transition of superconducting to normal conduction [6]. In addition, there are several theories that explain the superconducting phenomena [7]. According to the application there are low-temperature (LTS) superconductors and high-temperature superconductors (HTS) are available [8]. The majority of existing applications as well as the business for the industry still rely on LTS [9]. The superconducting properties of materials are also playing an important role in the field of research activities. As an example, the large magnetic system is used in the LHC for guiding the particles to keep in its trajectory [10]. Besides that, the high-field Nuclear Magnetic Resonance (NMR) devices are nearing 900 MHz and 1000MHz are equivalent to 21.2 T and 23.5 T in the magnetic bore, is the most difficult to the modern LTSs and as well as HTSs technology [11]. Trends for HTS utilization, notably in low field open system still thriving the MRI magnet market. The discovery of HTS piqued the interest of commercial applications involving such materials. The discovery of high temperature superconductors introduces a new pathway in the research field of superconductors. The high temperature super conductor like ceramic cuprates easily exceeds the critical temperature of nitrogen and it supports the further applications of superconductors. Another high temperature superconductor, MgB_2 is discovered in 2001 by [12] having T_c of 39K. Cuprate HTS material was discovered in 1986 and has since gained widespread attention due to its ability to eliminate the time-consuming and inefficient cooling device methods that were previously required. Although the superconducting material was discovered it took sometimes to further discovery of different superconducting materials suitable for the power applications. In fact, at the beginning of the 1911 superconducting materials are greatly used in the technological field, which includes transmission power cables, energy storage devices, superconducting magnets and many other applications.

2.   Meissner effect: Attribute to superconductors

In the year of 1933, the German Physicians Robert Oschsenfeld and Walther Meissner investigate that superconductors exhibit different characteristics from traditional perfect conductors. The investigation reveals that at certain temperature and magnetic field value, the magnetic field lines do not penetrate inside the superconductors and referred as Meissner effect (shown in Fig. 1) [13]. This phenomenon concludes that superconductors are diamagnetic with almost zero resistivity. Such absence of energy dissipation leads to flow of super current in the conducting circuit [14]. This discovery it is identified that superconductors are diamagnetic material and identified as new state of matter having special characteristics of zero resistivity with negligible power dissipation [14, 15].

There are two types of superconductors are available on the basis of response of the material in presence of external magnetic field, Type I and Type II superconductors [16]. Some metallic alloys and materials have electrical conductivity in room temperature, behaves as superconductor in low ambient temperature. The first discovered superconducting material Hg with zero resistivity at 4.2K is the Type I superconducting material [17]. The Type II superconductors are also formed with metals and alloys. There are some pure metallic superconductors are existing Niobium (Nb), Vanadium (V), carbon (C) and Technetium (Tc) is Type II superconductors [18].

The discovery of HTSs materials makes possible the superconducting devices to operate at much higher temperature than the liquid nitrogen and which is much easier to operate the superconducting devices [19]. Moreover, the high current density and higher critical magnetic field is also the crucial parameters along with the high transition temperature (T_c) [19, 20].

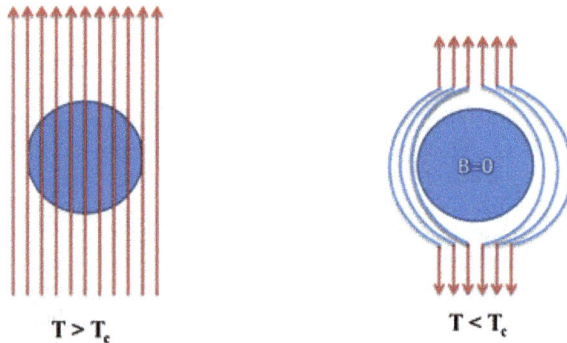

Fig. 1. Meissner effect.

3.   Advanced power transmission system

In the electrical transmission system, the use of superconducting material is one of the advantageous applications in the recent trends of engineering applications. The use of superconducting materials reduces the resistive losses and provides high current densities for high power applications [21]. This is a very crucial material science problem, but HTS cables of this type are currently available and have also been validated in power systems, which is a good thing. The HTS cables are as long as required and highly flexible with desired physical properties for used in commercial purpose. Such cables are also used in power transmission systems of synchronous condenser, motors, transformers, and generators [21]. The world largest HTS cable was used in 2014 in the power grid of Essen, Germany. The one kilometre 10kV HTS cables are connected between two substations and cooled by liquid nitrogen [22]. This HTS cables are more efficient than any other conventional cables that transport five times more electricity with almost no loss [23].

Another application in the electrical transmission of superconducting materials is the fault current limiters. The new technology of HTS materials is also capable to withstands the fault currents. Short-circuit currents are limited by such superconducting device, which are dependent on the utilities transmission and distribution networking systems that makes use of these HTS [22, 23]. When the current in HTS device exceeds a certain threshold value, it behaves like a typical resistive conductor, switching from high to low current as necessary. Due to such interesting features the HTS materials are now being extensively studied to apply in different field of applications. Electrical power transmission system is shown in Fig. 2.

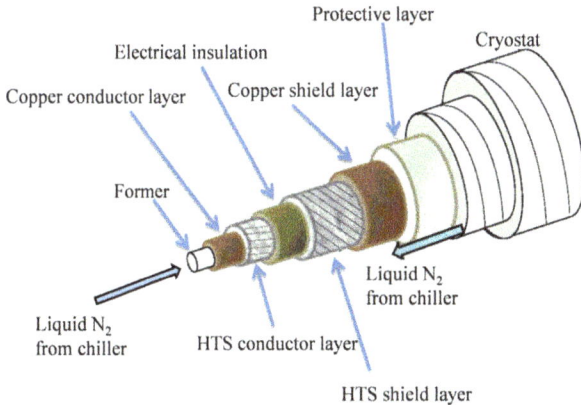

Fig. 2. Power cable transmission system.

4.   Super conducting electrical power devices

Apart from the electrical transmission, the superconducting technology is also used in electrical power devices. The appearance of liquid helium (LHe) cooling based low temperature superconductor and their usage in transformer windings sparked the interest of superconductivity in superconductors. Indeed, several manufacturers from around the world, including Europeans, Americans (Westinghouse) and Japanese (KEPC) explored these devices. However, uneconomical low-temperature superconductor devices and high cost of LHe, cooling systems made it more difficult for all nations to finance. The discovery of HTS materials has piqued people's interest as it eliminates the conventional cumbersome cooling devices [24]. HTS transformers have several advantages with higher efficient working ability and it can operate above rated power without reducing transformers life [25]. A three-phase transformer of 630 kV was tested by ABB company, which can use over a year under normal operating conditions. In comparison to the conventional devices, superconducting transformers made of HTS materials have the potential to provide significant economic, operational and environmental benefits. Siemens Company developed a transformer of 1MVA for the railway applications [26]. Furthermore, by eliminating the mineral oil of refrigeration, HTS transformers reduce the risk of fire according to the guidelines of EU. HTS operating at 77K promise to replace the copper windings of transformers and reducing the losses up to 70-80% [26].

Rotating electromechanical machines like motors and generators are the advanced application of superconducting materials. In general, the windings of rotor and the armature bars are associated with losses of generators. In case of large AC generators, the applications of superconducting technology for windings of rotor provides virtually unlimited field of capability with very resistive losses in the windings [18]. The development of large superconducting generators will necessitate a significant, long-term effort to produce a machine with performance and reliability comparable to conventional generators. The HTS enable generators can provide quick response and reactive power support. AS an example, Siemens company developed a propulsion motor of 4MW, 1500rpm with slot less armature windings of HTS material [27]. Moreover, the increased efficiency, improved capability, potentials for higher voltage machines, enhanced steady state, decrease in size and weight reduce outside generator constructions are all potential benefits.

Copper coils are often used in rotating electromechanical devices; however, these are being substituted with coils composed of HTS materials, which are mostly used in rotor coils and it provides several advantages rather than the conventional designs. Introduction of such HTS system reduces the losses and the rotor coils of HTS materials generate the higher magnetic fields. By substituting superconducting wires for traditional copper coils, it is possible to construct motors which are eco-friendly, smaller and lighter in weight as well as more affordable while also eliminating a significant source of motor noise. Some machines have been developed as a result of the benefits of using this technology in motors. In the year of 2009, the American Superconductor Corp. and the Northrop Grumman Corp. created a 36.5MW high-temperature superconductor ship propulsion motor using HTS materials [28]. As a result of using HTS materials as coils, the current capacity of the system increases to 150 times than the similar size of copper wire and the reduces the size of the along with ship weight by near about 200 metric tons [29]. In the other side, Sumitomo Electric and a group of Japanese researchers has successfully developed a HTS motor for ship of 4.4 ton having power of 365kW [30]. They also developed a superconducting motor for electrical passenger car of 30kW with 120Nm torque [30]. These examples demonstrated that superconducting motors will allow for higher torque, higher energy efficiency and less space.

Superconductors - Materials and Applications   Materials Research Forum LLC
Materials Research Foundations 132 (2022) 79-96   https://doi.org/10.21741/9781644902110-4

5. Advanced power storage system

The storing of energy is very challenging and essential for the modern scientific experiments and usages. In recent times to increases the capacity of the energy storing system superconducting technologies are used on the basis of magnetic energy storage. In reality, these Superconducting Magnetic Energy Storage (SMES) systems are essentially a cryogenically cooled superconducting coil inside which electrical energy is preserved as a frictionless flowing electric super current [31]. These SMES devices are characterised by its ability to store and discharge enormous amount of electricity in an almost instantaneous manner. As a result, they play an important role in transmission and distribution in electrical system. These systems also have the capability of power quality improvement of sensitive loads while having a lesser environmental impact than alternative storage devices such as batteries and have long lifetime. Flywheel Energy Storage (FES) devices are another advantageous application of superconductivity [32]. In case of FES devices, a bearing is used that is accelerated to extremely high speeds while maintaining the energy conservation law in turns of rotational energy. The storage is proportional to both mass of the moving body and square of the rotational speed. The spinning speed of the flywheel is lowered or increased while the energy is taken or added from the system according to the conservation of energy. Superconducting magnetic bearings (SMBs) suspended a flywheel disc connected to the bulk superconductors without any physical contact. The short-term energy storage flywheels with HTS bearing have been extensively studied in terms of technical and economic feasibility. Transition Temperature of some superconductors is shown in Table 1.

Table 1. Transition Temperature of some superconductors

Compound	Transition temperature (T_c) in K	Ref.
$SnSe_2(CO(C_5H_5)_2)_{0.33}$	6.2	[33]
Ba_8Si_{46}	8.09	[33]
$C_6Li_3Ca_2$	12.16	[34]
$PbMo_6S_8$	11.61	[33]
$Nd_{1.85}Ce_{0.15}CuO_4$	25	[35]
K_3C_{60}	19.6	[34]
$Ba_{0.6}K_{0.4}BiO_3$	32	[35]
FeSe	35.8	[36]
$SrFe_2As_2$	41	[35]
$YBa_2Cu_3O_7$	95	[36]
$BiSrCaCu_2O_{6+y}$	106	[37]
$HgBa_2CaCu_2O_{6+y}$	122	[37]

Superconductors - Materials and Applications   Materials Research Forum LLC
Materials Research Foundations 132 (2022) 79-96   https://doi.org/10.21741/9781644902110-4

6.   Modern transportation

The central Japan Railway company and the Company's Railway Technical Research Institute have developed superconducting magnetic levitation trains using HTS technology, which is a very interesting application of HTS technology. An electrodynamics suspension system (EDS) is used in the superconducting Maglev (SCMaglev) trains [38]. The concept is built on the mechanism of magnetic repulsion between tracks and bogies. Superconducting conducting magnets are installed in the bogies of trains and metallic coils are placed in the guide ways. As a result, the system between the train and the rail becomes frictionless. In the year of 2004, the SCMaglev train was initiated in Shanghai (China) for commercial purpose [39]. It was attending450km/h in just 8 minutes of travel time [40]. The Japanese rail operator claimed to achieve the speed of 603km/h SCMaglev train on 48.2km magnetic-levitation test track and is expected to have in operation in 2027. It will travel 284km route with expected time of 40 minutes which is less than half of the time travelled in bullet train [40, 41].

7.   Advanced accelerators

In particle accelerators like LHC (Large Hadron Collider) at CERN, to keep the particles in their fixed orbit strong magnetic fields are required [42]. For the acceleration of higher energetic particles, enough stronger magnetic fields are required to fix the trajectory [42]. But the energy of the circular orbital accelerators is limited by the power of used magnets. The innovation of superconductivity and its technology provides a new shape to overcome such limitations. The huge magnetic fields to keep the particles in circular orbit are only provided by the superconductors [43].

8.   Magnetic resonance devices

Nuclear magnetic resonance spectroscopy (NMR) is an ideal analytical technique for non-invasive, non-restrictive, and quantitative analyses of basic and advance investigation physics, Material science, Medicine, Chemistry, Biology, and also in other fields.

8.1   Magnetic resonance imaging for medical diagnostics

One of the advantageous applications of superconductivity in medical science is magnetic resonance imaging (MRI). Researchers and pharmaceutical businesses employ NMR scientific tools for drug research and development [44]. Moreover, the images obtained from NMR plays an important role for medical diagnostics purpose in the developing field of medical science [45]. The NMR technology used the strong magnetic field capable of influencing the orientation of atomic nuclei that make the human body. The fat molecules and the protons of hydrogen atoms found in the water considered as tiny magnets [46]. As the human body comes into the magnetic field of the NMR device, these tiny magnets are oriented by the magnetic field. When the field is terminated, these tiny magnets are returned to its equilibrium position by emitting an electrical signal. This electrical signal helps to determine tissue types. In case of fat tissues, due to higher density the relaxation time of the proton is shorter and as a result dark spot will be obtained from the imaging. In reverse order, for lower density clear spots may be visible from the image diagnosis process.

Diagnosis using these images has now become a standard medical procedure. The annual sales of MRI systems are rapidly increases [47]. The use of HTS materials increases the resolution of the imaging system and its lower cost makes it available for the peoples [48]. Oxford Instrument and Siemens collaborated to build the first experimental system with HTS soils in 1998 [48]. This is an open framework with C-shaped construction with 0.2 T fields utilizing two cryogen free curls of Bi wires and cooled by utilizing single stage GM cryocooler [49]. This framework exhibits the specialized practicality of utilizing HTS materials in loops in MRI gadgets in the lower field level [50].

8.2   NMR spectroscopy

NMR spectrometers are a little however quickly developing business sector portion for superconductivity. In view of their high goal, they have become progressively significant in science and science research. Up to this point, frameworks with frequencies up to 850 MHz, relating to 18.9 T in the magnet bore, have been accessible and selling great [51]. Frameworks for 950 MHz, which implies 22.25 T in the magnet bore, are approaching culmination and may address as far as possible for which LTS wires are satisfactory. Endeavors are being made at NRIM, NHMFL, and our research facility, as a team with significant organizations, to grow even 1000 MHz frameworks (23.5 T in the drag) [51]. For this situation, the uplifting properties of HTS conductors at low temperatures should help with accomplishing the necessary extremely high attractive field. Without a doubt, current endeavors with Bi (2212) and Bi (2223) multifilament tape conductors seem promising. Many little curls as supplements to high handle magnets have effectively been explored, especially in Japan and the United States. The latest distributed outcomes in such manner come from NHMFL, where a 3 T concentric loop framework with a 16 mm internal bore gave 21 T in a 19 T foundation field, and NRIM, where a curl with a 48 mm bore gave 24.45 T in a 22.8 T foundation field [52]. In any case, the way from these magnificent outcomes to full-scale NMR loops stays troublesome. This is because of the rigid prerequisites for NMR magnets as far as field homogeneity and time consistency. The subsequent condition, specifically, could be definitive. The prerequisite for the time consistency of the attractive field B_0 for high goal NMR spectroscopy is $B/B_0 < 1/10^8$ h, which denotes the winding's resistivity, including superconducting joints and worked in steady mode should be of the request for 12 to15 Ω [52]. Such qualities have not yet been accomplished with HTS wires, and it is as yet muddled whether they can be accomplished at all with current HTS conduits.

8.3   Fast field cycle relaxometer

A fast field cycling (FFC) nuclear magnetic resonance relaxometer is a low field magnetic resonance developed on the basis of magnet connected to the power supply and a RF circuit and operating in the range of few kHz to 100MHz according to the mechanical construction

Superconductors - Materials and Applications   Materials Research Forum LLC
Materials Research Foundations 132 (2022) 79-96   https://doi.org/10.21741/9781644902110-4

[53]. The low field NMR technique is used in FFC NMR Relaxometer that evaluates the of longitudinal spin relaxation rate [52]. Using the FFC NMR the thermal activation energy and distinction and identification of different molecular system can be possible. Moreover, it has contribution in the separation of crystalline and amorphous. It is possible to identification of the presence of parametric substance using the NMR device [54]. FFC NMR quantify the nuclear spin lattice relaxation constant depending on the strength of magnetic field and come up with the local molecular dynamics across a broad frequency range [55, 56]. This helps to determine the complex biomolecules aggregation states like proteins.

In the other side, Helmoltz coils are developed for the magnetic induction purpose. Different configurations with distinct characteristics coils are developed based on the different purpose of magnetic induction [57]. Furthermore, a high current is required to reach the target magnetic induction using the FFC technique [58]. The high current leads to Joule heating effect in the coils, necessitating the use of complex cooling system [59]. A magnetic circuit was developed to overcome this issue, reduce the magnet current and increase the efficiency with reduced volume and weight of the Relaxometer [60]. The diamagnetic superconducting blocks of Bi-2223 were introduced in order to improve the magnetic induction homogeneity with minimum losses in the circuit [61]. This technological solution decreases the magnetic resistance in presence of liquid nitrogen and permits to shrink the losses due to Joule heating effect in the magnets with desired magnetic induction [62, 63].

9. SQUID

The superconducting quantum interference device (SQUID) of superconducting loops with two parallel Josephson junctions is used to measure delicate magnetic field [64, 65]. This device arranged as magnetometer to measure the extremely small magnetic fields and also able to measure enough small magnetic fields in the living organisms [66]. The magnetic field in the mouse brains were measured using SQUID to see if there was enough magnetism to credit their navigational skills to an internal compass. SQUIDs can detect the fields as low as 10^{-14} T and 3fT $Hz^{-1/2}$ of noise levels [67]. Pure led alloy/niobium containing 10% gold/indium is the typical superconducting materials for SQUID, as pure lead as its temperature is altered frequently. To keep superconductivity, the gadget must be kept at a temperature of few degrees below the absolute zero and cooled with liquid helium. The discovery of HTS demonstrates a new direction in the field of technology. Although LTS SQUID are more sensitive than HTS SQUID, but adequate for wide range of applications and more economic [68].

SQUIDs are particularly well suited for biological research for their great sensitivity. For instance, magneto encephalography (MEG) is a technique that uses readings from an array of SQUIDs to interpret cerebral activity within brains [69]. In part as SQUIDs can acquire signals with an acquisition rates that are several times faster than the maximum temporal frequency of interest in the signals released by the brain (kHz), MEG is capable of achieving excellent temporal resolution [70]. Magneto-gastricography is another application for SQUIDs, which is focused with capturing the faint magnetic fields produced by the stomach [71]. The magnetic marker monitoring approach, that can be used to track the movement of orally administered medications, is a unique implementation of SQUIDs that has not yet been explored [72, 73]. Specifically, SQUIDs have been used in interventional cardiology for magnetic field imaging (MFI) that is a technique which detects and measures the magnetic field produced by the heart to aid in diagnostics and risk assessment [74, 75]. Several applications of Transmission of power in electrical system using superconductors are shown in Table 2.

Table 2. Some applications of Transmission of power in electrical system using superconductors.

Year	Place	Length [m]	Capacity [MVA]	Ref.
2015	Japan (Yokohama)	250	200(66kV AC,5 kA)	[33]
2015	Corea (Jeju Island)	1000	154(154Kvac,3.75 kA)	[33]
2014	Japan (Ishikan)	2000	100(710kV DC, 5kA)	[76, 77]
2014	USA (Westchester)	170	96(13.8kVAC/4k A)	[78]
2013	Corea (Icheon)	100	154 (154kV AC, 3.75kA)	[78]
2011	China	360	13(1.3kV DC,10kA)	[77]
2008	USA(Long Island)	600	574(138kV AC,2.4kA)	[78]

Conclusion

The largescale application of superconducting material still depends on application of multifilament wire of Nb_3Sn and NbTi. In the recent years the wires are developed to reduce the ac losses and Joule losses. Framing of superconducting critical temperature, the type and range of industrial applications has been increasing. The complicated circuit arrangements with industrial benefits can be achieved by replacing superconducting material. The main challenges are to achieve the room temperature values of critical temperature of superconducting material, and it also includes that to appear at a material of easy employable. Further essential requirement is to manage the manufacturing cost, manufacturing process, flexibility and stability of the superconducting materials. To achieve all of the required qualities, the research in the field of superconductivity requires novel approaches and mindsets that go beyond our current knowledge. The core applications of superconducting material include the magnetic resonance, transportation of electrical power; Maglev train and superconducting magnetic energy storage process take the advantages of superconducting properties. Moreover, MRI covers the major part of the recent superconductors market and also still growing further. In may be possible in future, the superconducting material would be present in most of our advantageous technical devices and surprising novel techniques would be possible.

References

[1] P.F. Dahl, Kamerlingh Onnes and the Discovery of Superconductivity: The Leyden Years, 1911-1914, Hist. Stud. Phys. Sci. 15 (1984) 1-37. https://doi.org/10.2307/27757541

[2] D. van Delft, P. Kes, The discovery of superconductivity, Phys. Today. 63 (2010) 38-43. https://doi.org/10.1063/1.3490499

[3] S.M. Anlage, The physics and applications of superconducting metamaterials, J. Opt. 13 (2011) 024001. https://doi.org/10.1088/2040-8978/13/2/024001

[4] H. Rogalla, P.H. Kes, 100 Years of Superconductivity, CRC Press, 2011: pp. 864. https://doi.org/10.1201/b11312

[5] C. Wang, D. Zhang, G. Fu, J.P. Wu, Analytical Study of the Holographic Superconductor from Higher Derivative Theory, Adv. High Energy Phys. 2020 (2020) 1-10. https://doi.org/10.1155/2020/5902473

[6] J.R. Hull, M. Murakami, Applications of bulk high-temperature Superconductors, Proc. IEEE. 92 (2004) 1705-1718. https://doi.org/10.1109/JPROC.2004.833796

[7] R.L. Fagaly, Superconducting quantum interference device instruments and applications, Rev. Sci. Instrum. 77 (2006) 101101. https://doi.org/10.1063/1.2354545

[8] M. Bäcker, Energy and superconductors - applications of high-temperature-superconductors, Zeitschrift Für Krist. 226 (2011). https://doi.org/10.1524/zkri.2011.1330

[9] P. Komarek, Advances in large scale applications of superconductors, Supercond. Sci. Technol. 13 (2000) 456-459. https://doi.org/10.1088/0953-2048/13/5/304

[10] L. Rossi, L. Bottura, Superconducting Magnets for Particle Accelerators, Rev. Accel. Sci. Technol. 05 (2012) 51-89. https://doi.org/10.1142/S1793626812300034

[11] R.G. Sharma, The Phenomenon of Superconductivity and Type II Superconductors, in: 2021: pp. 15-72. https://doi.org/10.1007/978-3-030-75672-7_2

[12] S.R. Shinde, S.B. Ogale, R.L. Greene, T. Venkatesan, P.C. Canfield, S.L. Bud'ko, G. Lapertot, C. Petrovic, Superconducting MgB2 thin films by pulsed laser deposition, Appl. Phys. Lett. 79 (2001) 227-229. https://doi.org/10.1063/1.1385186

[13] V.G. Prokhorov, V.L. Svetchnikov, J.S. Park, G.H. Kim, Y.P. Lee, J.-H. Kang, V.A. Khokhlov, P. Mikheenko, Flux pinning and the paramagnetic Meissner effect in MgB 2 with TiO 2 inclusions, Supercond. Sci. Technol. 22 (2009) 045027. https://doi.org/10.1088/0953-2048/22/4/045027

[14] Q. Wang, Z. Ni, C. Cui, Superconducting Magnet Technology and Applications, in: Supercond. - Mater. Prop. Appl., InTech, 2012. https://doi.org/10.5772/48465

[15] S. Deng, C. Felser, J. Köhler, A Reverse Approach to Superconductivity, J. Mod. Phys. 04 (2013) 10-13. https://doi.org/10.4236/jmp.2013.46A003

[16] B. Rosenstein, D. Li, Ginzburg-Landau theory of type II superconductors in magnetic field, Rev. Mod. Phys. 82 (2010) 109-168. https://doi.org/10.1103/RevModPhys.82.109

[17] A. Martinelli, F. Bernardini, S. Massidda, The phase diagrams of iron-based superconductors: Theory and experiments, Comptes Rendus Phys. 17 (2016) 5-35. https://doi.org/10.1016/j.crhy.2015.06.001

[18] A. Roque, D.M. Sousa, V. Fernão Pires, E. Margato, Superconductivity and their Applications, Renew. Energy Power Qual. J. 1 (2017) 322-327. https://doi.org/10.24084/repqj15.308

[19] J.R. Hull, Applications of high-temperature superconductors in power technology, Reports Prog. Phys. 66 (2003) 1865-1886. https://doi.org/10.1088/0034-4885/66/11/R01

[20] D. Larbalestier, A. Gurevich, D.M. Feldmann, A. Polyanskii, High-T c superconducting materials for electric power applications, in: Mater. Sustain. Energy, Co-Published with Macmillan Publishers Ltd, UK, 2010: pp. 311-320. https://doi.org/10.1142/9789814317665_0046

[21] M. Noe, M. Steurer, High-temperature superconductor fault current limiters: concepts, applications, and development status, Supercond. Sci. Technol. 20 (2007) R15-R29. https://doi.org/10.1088/0953-2048/20/3/R01

[22] M.W. Rupich, Second-generation (2G) coated high-temperature superconducting cables and wires for power grid applications, in: Supercond. Power Grid, Elsevier, 2015: pp. 97-130. https://doi.org/10.1016/B978-1-78242-029-3.00004-2

[23] M. Ikram, A. Raza, S. Altaf, A. Ahmed Rafi, M. Naz, S. Ali, S. Ossama Ali Ahmad, A. Khalid, S. Ali, J. Haider, High Temperature Superconductors, in: Transit. Met. Compd. - Synth. Prop. Appl., IntechOpen, 2021. https://doi.org/10.5772/intechopen.96419

[24] L. Shao, M. Ehrgott, An approximation algorithm for convex multiplicative programming problems, in: 2011 IEEE Symp. Comput. Intell. Multicriteria Decis., IEEE, 2011: pp. 175-181. https://doi.org/10.1109/SMDCM.2011.5949275

[25] R. Schlosser, H. Schmidt, M. Leghissa, M. Meinert, Development of high-temperature superconducting transformers for railway applications, IEEE Trans. Appiled Supercond. 13 (2003) 2325-2330. https://doi.org/10.1109/TASC.2003.813118

[26] S. Mishra, T.A. Lipo, S. V. Pamidi, Design and analysis of a novel brushlesshigh temperature superconducting synchronous machine, in: 2017 IEEE Int. Electr. Mach. Drives Conf., IEEE, 2017: pp. 1-6. https://doi.org/10.1109/IEMDC.2017.8002183

[27] K. Ilieva, O. Dinolov, State-of-the-art of superconducting materials and their energy-efficiency applications, in: 2020 7th Int. Conf. Energy Effic. Agric. Eng., IEEE, 2020: pp. 1-5. https://doi.org/10.1109/EEAE49144.2020.9279004

[28] X. Li, Superconducting Devices in Wind Farm, in: Wind Energy Manag. 2011: pp. 140. https://doi.org/10.5772/17969

[29] S. Mishra, T.A. Lipo, S. V. Pamidi, Design and analysis of a novel brushlesshigh temperature superconducting synchronous machine, in: 2017 IEEE Int. Electr. Mach. Drives Conf., IEEE, 2017: pp. 1-6. https://doi.org/10.1109/IEMDC.2017.8002183

[30] X. Luo, J. Wang, M. Dooner, J. Clarke, Overview of current development in electrical energy storage technologies and the application potential in power system operation, Appl. Energy. 137 (2015) 511-536. https://doi.org/10.1016/j.apenergy.2014.09.081

[31] P.F. Ribeiro, B.K. Johnson, M.L. Crow, A. Arsoy, Y. Liu, Energy storage systems for advanced power applications, Proc. IEEE. 89 (2001) 1744-1756. https://doi.org/10.1109/5.975900

[32] H. Yaghoubi, The Most Important Maglev Applications, J. Eng. 2013 (2013) 1-19. https://doi.org/10.1155/2013/537986

[33] A.B. Holder, H. Keller, High-temperature superconductors: underlying physics and application, Z. Naturforsch. B. 75 (2020) 3-14. https://doi.org/10.1515/znb-2019-0103

[34] M. Suenaga, Metallurgy of Continuous Filamentary A15 Superconductors in Materials Science, in: Plenum Press, Springer, New York, London, 68 (1981). https://doi.org/10.1007/978-1-4757-0037-4_4

[35] J.G. Bednorz, K.A. Müller, Possible high T c superconductivity in the Ba−La−Cu−O system, Z. Phys. B Condens. Matter. 64 (1986) 189-193. https://doi.org/10.1007/BF01303701

[36] P.N. Barnes, J.W. Kell, B.C. Harrison, T.J. Haugan, Minute doping with deleterious rare earths in YBa2Cu3O7− YBa2Cu3O7−δ films for flux pinning enhancements, Appl. Phys. Lett. 89 (2006) 012503. https://doi.org/10.1063/1.2219391

[37] S.P. Singh, R.K. Pandey, P. Singh, Cooper pair breaking and isotope effect coefficient variation in high-T c superconductors, J. Supercond. 9 (1996) 269-271. https://doi.org/10.1007/BF00727546

[38] H.S. Han, B.H. Yim, N.J. Lee, Y.J. Kim, Prediction of ride quality of a Maglev vehicle using a fullvehicle multi-body dynamic model, Veh. Syst. Dyn. 47 (2009) 1271-1286. https://doi.org/10.1080/00423110802632063

[39] Z. Wang, X. Li, Y. Xie, Z. Long, Maglev Train Signal Processing Architecture Based on Nonlinear Discrete Tracking Differentiator, Sensors. 18 (2018) 1697. https://doi.org/10.3390/s18061697

[40] H. Almujibah, J. Preston, The total social costs of constructing and operating a maglev line using a case study of the riyadh-dammam corridor, Saudi Arabia, Transp. Syst. Technol. 4 (2018) 298-327. https://doi.org/10.17816/transsyst201843s1298-327

[41] S. Rao, M. Brüggen, J. Liske, Detection of gravitational waves in circular particle accelerators, Phys. Rev. D. 102 (2020) 122006. https://doi.org/10.1103/PhysRevD.102.122006

[42] S.A. Gourlay, Superconducting accelerator magnet technology in the 21st century: A new paradigm on the horizon?, Nucl. Instruments Methods Phys. Res. Sect. A Accel. Spectrometers, Detect. Assoc. Equip. 893 (2018) 124-137. https://doi.org/10.1016/j.nima.2018.03.004

[43] E. Moser, E. Laistler, F. Schmitt, G. Kontaxis, Ultra-High Field NMR and MRI-The Role of Magnet Technology to Increase Sensitivity and Specificity, Front. Phys. 5 (2017). https://doi.org/10.3389/fphy.2017.00033

[44] C.M. Quinn, M. Wang, T. Polenova, NMR of Macromolecular Assemblies and Machines at 1 GHz and Beyond: New Transformative Opportunities for Molecular Structural Biology, in: Protein NMR, 2018: pp. 1-35. https://doi.org/10.1007/978-1-4939-7386-6_1

[45] W. Braun, G. Wider, K.H. Lee, K. Wüthrich, Conformation of Glucagon in a Lipid-Water Interphase by 1H Nuclear Magnetic Resonance, in: NMR in Structural Biology, 1995: pp. 264-291. https://doi.org/10.1142/9789812795830_0022

[46] R.L. Bernays, E.R. Laws, Intraoperative Diagnostic and Interventional Magnetic Resonance Imaging in Neurosurgery, Neurosurgery. 41 (1997) 999-999. https://doi.org/10.1097/00006123-199710000-00068

[47] B. Hu, K. Wang, L. Wu, S.-H. Yu, M. Antonietti, M.-M. Titirici, Engineering Carbon Materials from the Hydrothermal Carbonization Process of Biomass, Adv. Mater. 22 (2010) 813-828. https://doi.org/10.1002/adma.200902812

[48] S. Schaffer, Physics Laboratories and the Victorian Country House, in: Mak. Sp. Sci., Palgrave Macmillan UK, London, 1998: pp. 149-180. https://doi.org/10.1007/978-1-349-26324-0_7

[49] L. Bertora, MRI Magnets based on MgB2, in: MgB2 Superconducting Wires, 2016: pp. 485-536. https://doi.org/10.1142/9789814725590_0018

[50] S.S. Kalsi, K. Weeber, H. Takesue, C. Lewis, H.-W. Neumueller, R.D. Blaugher, Development status of rotating machines employing superconducting field windings, Proc. IEEE. 92 (2004) 1688-1704. https://doi.org/10.1109/JPROC.2004.833676

[51] S. Anders, M.G. Blamire, F.-I. Buchholz, D.-G. Crété, R. Cristiano, P. Febvre, L. Fritzsch, A. Herr, E. Il'ichev, J. Kohlmann, J. Kunert, H.-G. Meyer, J. Niemeyer, T. Ortlepp, H. Rogalla, T. Schurig, M. Siegel, R. Stolz, E. Tarte, H.J.M. ter Brake, H. Toepfer, J.-C. Villegier, A.M. Zagoskin, A.B. Zorin, European roadmap on superconductive electronics - status and perspectives, Phys. C Supercond. 470 (2010) 2079-2126. https://doi.org/10.1016/j.physc.2010.07.005

[52] A.E. Berns, S. Bubici, C. De Pasquale, G. Alonzo, P. Conte, Applicability of solid state fast field cycling NMR relaxometry in understanding relaxation properties of leaves and leaf-litters, Org. Geochem. 42 (2011) 978-984. https://doi.org/10.1016/j.orggeochem.2011.04.006

[53] D.M. Sousa, G.D. Marques, P.J. Sebastião, A.C. Ribeiro, New isolated gate bipolar transistor two-quadrant chopper power supply for a fast field cycling nuclear magnetic resonance spectrometer, Rev. Sci. Instrum. 74 (2003) 4521-4528. https://doi.org/10.1063/1.1610785

[54] P. Conte, V. Ferro, Measuring hydrological connectivity inside a soil by low field nuclear magnetic resonance relaxometry, Hydrol. Process. 32 (2018) 93-101. https://doi.org/10.1002/hyp.11401

[55] G. Parigi, E. Ravera, M. Fragai, C. Luchinat, Unveiling protein dynamics in solution with field-cycling NMR relaxometry, Prog. Nucl. Magn. Reson. Spectrosc. 124-125 (2021) 85-98. https://doi.org/10.1016/j.pnmrs.2021.05.001

[56] N. Abhyankar, V. Szalai, Challenges and Advances in the Application of Dynamic Nuclear Polarization to Liquid-State NMR Spectroscopy, J. Phys. Chem. B. 125 (2021) 5171-5190. https://doi.org/10.1021/acs.jpcb.0c10937

[57] Z. Yang, L. Zhang, Magnetic Actuation Systems for Miniature Robots: A Review, Adv. Intell. Syst. 2 (2020) 2000082. https://doi.org/10.1002/aisy.202000082

Materials Research Forum LLC
https://doi.org/10.21741/9781644902110-4

[58] K. Zhu, Y. Ju, J. Xu, Z. Yang, S. Gao, Y. Hou, Magnetic Nanomaterials: Chemical Design, Synthesis, and Potential Applications, Acc. Chem. Res. 51 (2018) 404-413. https://doi.org/10.1021/acs.accounts.7b00407

[59] F. Najmabadi, Spherical torus concept as power plants-the ARIES-ST study, Fusion Eng. Des. 65 (2003) 143-164. https://doi.org/10.1016/S0920-3796(02)00302-2

[60] W.K. Peng, L. Chen, J. Han, Development of miniaturized, portable magnetic resonance relaxometry system for point-of-care medical diagnosis, Rev. Sci. Instrum. 83 (2012) 095115. https://doi.org/10.1063/1.4754296

[61] A. Roque, D.M. Sousa, E. Margato, V. Malo Machado, P.J. Sebastiao, G.D. Marques, Magnetic Flux Density Distribution in the Air Gap of a Ferromagnetic Core With Superconducting Blocks: Three-Dimensional Analysis and Experimental NMR Results, IEEE Trans. Appl. Supercond. 25 (2015) 1-9. https://doi.org/10.1109/TASC.2015.2483599

[62] W.V. Hassenzahl, D.W. Hazelton, B.K. Johnson, P. Komarek, M. Noe, C.T. Reis, Electric power applications of superconductivity, Proc. IEEE. 92 (2004) 1655-1674. https://doi.org/10.1109/JPROC.2004.833674

[63] Z.S. Hartwig, C.B. Haakonsen, R.T. Mumgaard, L. Bromberg, An initial study of demountable high-temperature superconducting toroidal field magnets for the Vulcan tokamak conceptual design, Fusion Eng. Des. 87 (2012) 201-214. https://doi.org/10.1016/j.fusengdes.2011.10.002

[64] B.I. Oladapo, S.A. Zahedi, S.C. Chaluvadi, S.S. Bollapalli, M. Ismail, Model design of a superconducting quantum interference device of magnetic field sensors for magnetocardiography, Biomed. Signal Process. Control. 46 (2018) 116-120. https://doi.org/10.1016/j.bspc.2018.07.007

[65] R. Kleiner, D. Koelle, F. Ludwig, J. Clarke, Superconducting quantum interference devices: State of the art and applications, Proc. IEEE. 92 (2004) 1534-1548. https://doi.org/10.1109/JPROC.2004.833655

[66] T. Schwarz, J. Nagel, R. Wölbing, M. Kemmler, R. Kleiner, D. Koelle, Low-Noise Nano Superconducting Quantum Interference Device Operating in Tesla Magnetic Fields, ACS Nano. 7 (2013) 844-850. https://doi.org/10.1021/nn305431c

[67] C. Pfeiffer, L.M. Andersen, D. Lundqvist, M. Hämäläinen, J.F. Schneiderman, R. Oostenveld, Localizing on-scalp MEG sensors using an array of magnetic dipole coils, PLoS One. 13 (2018) 0191111. https://doi.org/10.1371/journal.pone.0191111

[68] A.I. Braginski, Superconducting electronics coming to market, IEEE Trans. Appiled Supercond. 9 (1999) 2825-2836. https://doi.org/10.1109/77.783621

[69] S. Braeutigam, Magnetoencephalography: Fundamentals and Established and Emerging Clinical Applications in Radiology, ISRN Radiol. 2013 (2013) 1-18. https://doi.org/10.5402/2013/529463

[70] M.K. Abadi, R. Subramanian, S.M. Kia, P. Avesani, I. Patras, N. Sebe, DECAF: MEG-Based Multimodal Database for Decoding Affective Physiological Responses, IEEE Trans. Affect. Comput. 6 (2015) 209-222. https://doi.org/10.1109/TAFFC.2015.2392932

[71] S. Baillet, J.C. Mosher, R.M. Leahy, Electromagnetic brain mapping, IEEE Signal Process. Mag. 18 (2001) 14-30. https://doi.org/10.1109/79.962275

[72] A. Sorriento, M.B. Porfido, S. Mazzoleni, G. Calvosa, M. Tenucci, G. Ciuti, P. Dario, Optical and Electromagnetic Tracking Systems for Biomedical Applications: A Critical Review on Potentialities and Limitations, IEEE Rev. Biomed. Eng. 13 (2020) 212-232. https://doi.org/10.1109/RBME.2019.2939091

[73] A.K.A. Silva, E.L. Silva, J.F. Carvalho, T.R.F. Pontes, R.P. de A. Neto, A. da S. Carriço, E.S.T. Egito, Drug Targeting and other Recent Applications of Magnetic Carriers in Therapeutics, Key Eng. Mater. 441 (2010) 357-378. https://doi.org/10.4028/www.scientific.net/KEM.441.357

[74] G. Lembke, S.N. Erné, H. Nowak, B. Menhorn, A. Pasquarelli, G. Bison, Optical multichannel room temperature magnetic field imaging system for clinical application, Biomed. Opt. Express. 5 (2014) 876. https://doi.org/10.1364/BOE.5.000876

[75] J. Chen, J. Yang, R. Liu, C. Qiao, Z. Lu, Y. Shi, Z. Fan, Z. Zhang, X. Zhang, Dual-targeting Theranostic System with Mimicking Apoptosis to Promote Myocardial Infarction Repair via Modulation of Macrophages, Theranostics. 7 (2017) 4149-4167. https://doi.org/10.7150/thno.21040

[76] S. Nishijima, S. Eckroad, A. Marian, K. Choi, W.S. Kim, M. Terai, Z. Deng, J. Zheng, J. Wang, K. Umemoto, J. Du, P. Febvre, S. Keenan, O. Mukhanov, L.D. Cooley, C.P. Foley, W.V. Hassenzahl, M. Izumi, Superconductivity and the environment: a roadmap, Supercond. Sci. Technol. 26 (2013) 113001. https://doi.org/10.1088/0953-2048/26/11/113001

[77] H. Thomas, A. Marian, A. Chervyakov, S. Stuckrad, D. Salmieri, C. Rubbia, Superconducting transmission lines - Sustainable electric energy transfer with higher public acceptance?, Renew. Sust. Energ. Rev. 55 (2014) 59-72. https://doi.org/10.1016/j.rser.2015.10.041

[78] T. Bohno, A. Tomioka, M. Imaizumi, Y. Sanuki, T. Yamamoto, Y. Yasukawa, H. Ono, Y. Yagi, K. Iwadate, Development of 66kV/6.9kV 2MVA prototype HTS power transformer, Physica C Supercond. 426-431(2005) 1402-1407. https://doi.org/10.1016/j.physc.2005.03.080

Superconductors - Materials and Applications
Materials Research Foundations 132 (2022) 97-107

Materials Research Forum LLC
https://doi.org/10.21741/9781644902110-5

Chapter 5

Lanthanide-based Superconductor and its Applications

Godlisten N. Shao[1]*,

[1]Department of Chemistry, Mkwawa College, University of Dar es Salaam, P.O. Box 2513, Iringa, Tanzania

godlisten.shao@muce.ac.tz or shaogod@gmail.com

Abstract

Superconductors are materials that conduct electricity with no resistance below its critical temperature (T_c). To date, pure metals, metal alloys, oxides, hydrides and super hydrides are among structures that have been reported to exhibit excellent superconducting properties due to their unique electronic properties and lattice structure. Most researchers have widely reported on the fabrication, structure, properties and applications of cuprate and iron-based superconducting materials. The modification of cuprate-based and iron-based superconducting materials using lanthanides have shown to massively improve their physico-chemical properties and applications. Investigations on lanthanide superhydride superconductors which contain hydrogen framework structures such as LaH_{10} and YbH_{10} are a recent adventure in the field of superconductors. Lanthanide-based structures are considered as potential high temperature superconductors (HTSC) and can be used in high performance applications. The current chapter outlines the advances and prospects observed in lanthanide-based superconductors (LBSC) as modern and fascinating functional materials. There is some literature that has been dedicated to providing a review on superconductors but very few have reported on LBSC. This review chapter provides a general insight of the development of LBSC and their potential technological applications.

Keywords

Superconductivity, Critical Temperature, High-Temperature Superconductors, Lanthanide-based Superconductors

Contents

1.   Introduction

Superconductors are useful functional materials in telecommunications, power transmission, computer technology, transportation and medicine [1-3]. Of the superconductors that have been investigated, high-temperature superconducting materials exhibit desirable properties that attract application in the manufacturing of electrical cables, magnetic resonance imaging [4], sensors for brain diagnostics [5], and in transportation [6]. Thus, their enormous importance has essentially prompted various researchers to work on these fascinating functional materials.

Despite the emergency of various theories that accounts for the mechanism of superconducting materials, the Bardeen-Cooper-Schrieffer (BCS) theory and Debye's model still serve as the best models to describe the mechanism of superconductivity in most of superconducting materials [1, 7, 8]. The superconductivity depends on the electronic properties and lattice structure of materials. The critical point in improving the superconductivity is to ensure the formation of materials with high critical temperature and high current density [9-11].

Most of researches were invested in studying cuprate-based superconductors (CBSC). However, nowadays the iron-based superconductors (IBSC) are very useful functional materials that have been widely investigated as well. Initially, it was thought that superconductivity can only occur at low temperatures [12]. Notwithstanding, a superconductivity at 92 K and 130 K was essentially observed in $YBa_2Cu_3O_x$ and $Hg_2Ba_2Ca_2Cu_3O_x$, respectively [13, 14]. These materials were considered as high-temperature superconductors (HTSC) due to their unique ability to be active at higher temperatures. Since that breakthrough, the quest for high T_c superconductors has eventually led to discovery of various classes of superconductors [1, 3, 12, 15]. With no time, most of researches seem to migrate from studying CBSC to FBSC. This started with investigating the derivatives of $LaFePO$ and $LaFeAsO_{1-x}F_x$ which possess T_c of 4 K and 26 K, respectively [16, 17]. Figure 1 presents a diagram of the tetragonal structure of LaO_1.

$_x$F$_x$FeAs and the functional temperature change with increasing the amount of F$^-$ in the microstructure. It has been reported [16-19] that introducing other rare earth elements in LaOFeAs such Ce, Sm, Nd, and Pm can lead to the formation of HTSC with desirable properties for various applications. Table 1 shows notable advances achieved in the formation of superconducting materials with various T$_c$. However, HTSC are known to be expensive materials and their preparation processes involve complicated synthetic routes. Hence, this drawback hampers large-scale production and therefore the practical applicability of the HTSC has not yet widely been realized. Besides, iron-based superconductors have shown a critical temperature higher than 100 K. In addition, lanthanide superhydrides have shown to possess critical temperatures $\leq$250 K. Interestingly, the mechanism of superconductivity in iron-based superconductors cannot be described by the BCS theory. This indicates that the mechanism of superconductivity should be embedded in diverse theories depending on the nature of the superconducting materials. Based on the aforementioned aspects, the current review suggests the superconducting materials such as CBSC and IBSC doped with lanthanides (rare earth metals) to be categorized as lanthanide-based superconductors (LBSC). Therefore, it is the role of this chapter to provide a review on the utilization of lanthanides in forming functional LBSC with unique properties for a myriad of technological applications.

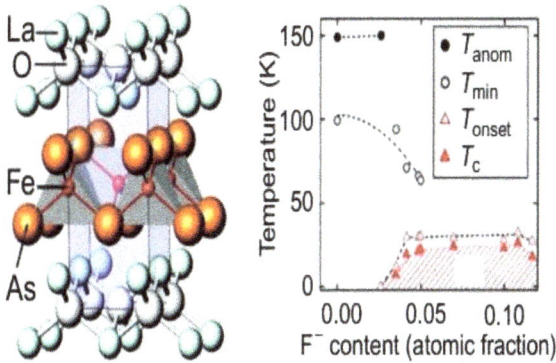

Figure 1: The left diagram shows the tetragonal structure of LaO$_{1-x}$F$_x$FeAs. The right plot shows the relationship between the changing in functional temperature with the amount of dopant (F$^-$ ions in atomic fraction). T$_{anom}$ denotes the temperature observed in undoped LaOFeAs as it decreases to the minimum value (T$_{min}$). The temperature at which the superconducting properties start is denoted as T$_{onset}$. Tc shown is for F-doped LaOFeAs [3, 17]

Superconductors - Materials and Applications   Materials Research Forum LLC
Materials Research Foundations 132 (2022) 97-107   https://doi.org/10.21741/9781644902110-5

Table 1: List of lanthanide element containing superconducting materials and their transition temperatures

Material	T_c	Reference
$CeFeAsO_{1-x}H_x$	48.0	[7]
$LaO_{0.89}F0_{.11}FeAs$	26.0	[8]
$LaO_{0.9}F_{0.2}FeAs$	29.0	[8]
$CeFeAsO_{0.84}F_{0.1}$	41.0	[8]
$Sr_{0.5}Sm_{0.5}FeAsF$	56.0	[8]
$La_{0.5}Y_{0.5}FeAsO_{0.6}$	43.1	[8]
$NdFeAsO_{0.89}F_{0.11}$	52.0	[8]
$PrFeAsO_{0.89}F_{0.11}$	52.0	[8]
$ErFeAsO_{(1-x)}$	45.0	[8]
LaH_{10}	260	[15]
$SmFeAsO_{0.9}F_{0.1}$	43.0	[16]
$SmFeAsO_{0.85}$	55.0	[16]
$LaFeAsO_{0.15-0.85}$	22.0	[27]
$LaFeAsO_{0.85}H_{(0-0.85)}$	35.0	[28]
$LaFeAsO_{1-x}H_x$	36.0	[29]
$SmFeAsO_{1-x}H_x$ $_{(x=0-0.85)}$	56.0	[34]
$NdFeAsO_{1-x}H_x$	54.0	[34]

2.   Lanthanide-based superconductors

The study by Bednorz and Muller [9, 20] in 1986 disclosed that the superconductivity at 30 K can be obtained in polycrystalline $LaBa_2CuO_{4-x}$. It has been reported that H_3S can exhibit superconducting properties at 203 K and pressure of 150 GPa [12]. This report has provided a gateway to fabrication of hydrogen rich superconductor systems with superior properties for various applications. This signifies that high-temperature superconductivity can be attained in materials with high phonon frequencies. The presence of filled or unfilled f-shells gives lanthanides unique electronic properties that contribute to their excellent superconducting properties. Doping the main classes of superconductors (cuprate-based superconductors and iron-based superconductors) with lanthanides leads to the formation of HTSC with appealing superconducting properties. The ability to generate functional materials with improved superconducting properties is of great importance towards the realization of the practical applications of superconductors. Table 1 shows the incorporation of lanthanides into various materials to generate superconductors with different critical temperatures. It can be observed that T_c depends on the type of the lanthanide element incorporated into the matrix to form superconducting materials.

2.1   Preparation methods

Most of the LBSC are HTSC since they can become superconductive at much higher temperatures. Customarily, the preparation of HTSC is believed to be difficult and too expensive for use over long distances. Hence, computational investigations have been

Superconductors - Materials and Applications   Materials Research Forum LLC
Materials Research Foundations 132 (2022) 97-107   https://doi.org/10.21741/9781644902110-5

widely used in predicting superconductors with desirable properties for a myriad of applications [18]. Superconductor materials can be synthesized using various methods such as solid state reaction processes, laser heating, molecular beam epitaxy pulsed laser deposition and high-pressure synthetic processes. The solid state reaction, laser heating and high-pressure methods are common synthetic routes used to prepare LBSC [16, 17, 21, 22].

2.1.1 Solid state reaction processes

According to the method given by Wang et al. [21] and Wu et al. [22], superconducting materials can be achieved using conventional solid state reaction processes. In a typical preparation process, precursors in the form of powders and at high purity (<99%) are sealed in evacuated quartz tubes, and heated at higher temperatures ($\leq$600 °C to 750 °C). Lanthanide oxides are dehydrated through preheating at 900 °C for 24 hours to obtain dry samples. The obtained materials are ground, pelletized, and eventually sintered at temperatures ranging from 940 °C to 970 °C [16, 17, 21-23]. The advantage of this method is that pure and stable final products can be yielded. However, the utilization of expensive equipment and high temperature requirements can hamper their large-scale production.

2.1.2 Laser heating

Laser heating and specifically laser-heated diamond-anvil cell is a technological method that facilitates economical measurement of physical properties of materials at high pressures and temperatures [15, 24]. This method is convenient in synthesizing hydrogen-rich compounds, XH_n with $n>5$ (for example, MgH_6, CaH_6, LaH_{10}, YH_6 etc.). Thus, lanthanide-based superconductors such as lanthanide superhydrides have been achieved through this method. The method allows tuning of pressure and chemical composition during chemical reaction. Lanthanides and H_2 are used as precursors at high pressures and temperatures [15, 24, 25].

2.1.3 High-pressure synthesis

This method involves synthesis of materials by solid state process under elevated pressures. The approach is very preferable in synthesizing equiatomic compounds of d and perhaps f orbitals in order to study their physical properties at low temperature conditions. It is very applicable in the preparation of LaIrP, LaIrAs and LaRhP superconductors [10, 26].

2.2 Characterization of lanthanide-based superconductors

Various characterization methods can be used to provide the general insight of the physico-chemical properties and practical applicability of the lanthanide-based superconductors. Thus, the electrical, lattice structure, chemical states and magnetic properties can be exquisitely investigated. The crystal structure of the samples can be studied using an X-ray diffractometer (XRD). It should be understood that the superconductivity of material depends on the lattice structure [1, 10, 17, 26]. A few to mention is the report by Wu et al.[22] that investigated the superconductivity at 33 - 37 K in $CsLn_2Fe_4As_4O_2$ and $KLn_2Fe_4As_4O_2$ (Ln = Lanthanides: Dy, Er, Gd, Ho,Nd, Sm and Tb). It was found that Tc

Superconductors - Materials and Applications   Materials Research Forum LLC
Materials Research Foundations 132 (2022) 97-107   https://doi.org/10.21741/9781644902110-5

correlates with the lattice structure especially for all 12442-type superconductors. It was further disclosed that Tc tends to be influenced by the lattice mismatch. Figure 2 shows a representative figure of the crystal structure and lattice parameters of 12442-type $CsLn_2Fe_4As_4O_2$, $RbLn_2Fe_4As_4O_2$ and $KLn_2Fe_4As_4O_2$ obtained from the intergrowth of 122-type K, Rb or $CsFe_2As_2$ and 1111-type LnFeAsO. These results reveal the dependence of the superconductivity on the lattice structure [27-29]. The data from crystal structure and the electronic configuration are very essential in estimating the superconducting transition temperatures of various HTSC superconductors using the Roeser–Huber equation [11].

Roeser–Huber equation for estimating the T_c

$$T_c = h^2/ [(2x)^2 2M_L] n^{-2/3} \pi k_B$$

Where, T_c, the superconducting transition temperature;

h, is the Planck's constant; x, the characteristic distance; M_L, the mass of the charge carriers; n, the correction factor relating the number of superconductor's planes in the unit cell; and k_B, is the Boltzmann constant.

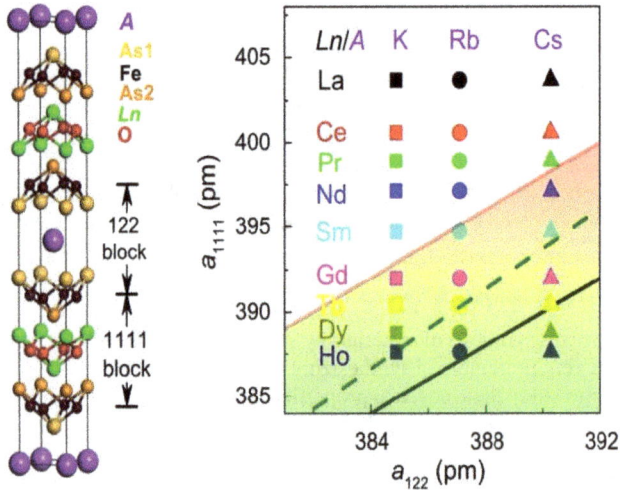

Figure 2. *Figure 2 (left) presents a 12442-type crystal structure for superconducting material, $MLn_2Fe_4As_4O_2$ (M = K, Rb, and Cs; Ln = Lanthanides), obtained from MFe_2As_2 (122-type) and LnFeAsO (1111-type). A plot showing the interacting lattice parameters between MFe_2As_2 and $LnFeAsO_{11}$ sources to obtain a 12442-type product is presented on the right panel. The proper lattice match is located in the lower and upper straight lines [21].*

The grain size has been reported to affect the superconductivity of materials as well. Microstructures of the superconductor materials can be examined by high resolution transmission electron microscopy (HRTEM) and scanning electron microscopy (SEM). SEM coupled with Energy Dispersive X-Ray Spectroscopy (EDX) allows the identification of elemental content since most superconducting materials contain several elements. The effect of grain boundary on the superconducting materials has been studied as well. Grain boundaries can block movement of supercurrent in superconductors thus optimization of conditions to obtain materials with desirable grain boundary is of paramount importance. Studies on the electrical properties of superconducting material are of great importance as well. The investigation of the electrical resistivity can be accomplished using a standard four-probe method on a Physical Property Measurement System (PPMS). The differential thermal analysis (DTA-TGA) can be employed to study the melt temperature. Magnetic measurements can be performed by a vibrating sample magnetometer using PPMS. Generally, most spectrometric methods can be used to study the microstructure of superconductors.

2.3 Superconducting properties of the LBSC

There is no a standalone theory that discloses the role of lanthanides in the superconducting materials. The influence of the lanthanides depends on the existing correlation between the lattice and electronic structures of a given superconductor [18, 19, 23, 25]. The superconducting properties of LBSC can be studied by considering the conductivity of lanthanide superhydrides. Since lanthanides possess electrons in the f orbital, the superconducting properties of materials are largely affected by the f states. The lanthanide-based superconductors possess highly localized f electrons which can affect their superconductivity [15, 24, 25]. It has been reported that LaH_{10} and CeH_9 can attain high T_c of 250-260 K and 117 K, respectively. The T_c below 10 K has been registered in PrH_9 and NdH_9 materials [19, 30]. It has been observed that the critical temperature for lanthanide hydrides decreases with increasing the number of electrons in the f-shell and superconductivity disappears for half-filled f-shells $(Eu–f^7)$ [30].

Interestingly, the superconducting properties of TmH_6, YbH_6 and LuH_6 are also influenced by the electrons present in the 4f orbitals as well. Tm in TmH_6 possessing 13 electrons in the 4f orbitals (unfilled) makes the compound stable at 50 GPa and has a low value of T_c (25 K). On the other hand, both YbH_6 and LuH_6 compounds have Yb and Lu elements which contain 14 electrons in their 4f orbitals. The decrease of the f-energy below the Fermi level is thought to cause formation of the nesting regions on the Fermi surface. Therefore, high critical temperature values registered in these compounds is due to phonon "softening". This role of f-orbital electrons in influencing the superconductivity has been further investigated by Song et al. [19]. The report investigated the influence of the f states on the Fermi surface, nesting and the value of T_c superconductivity in heavy rare earth hydrides. Drozdov et al. [12] investigated the superconductivity at 250 K in the lanthanum hydride system under high pressures. Moreover, Shipley et al [31] reported on the stability and superconductivity of lanthanum and yttrium decahydrides. All of these reports demonstrate that lanthanides are very essential superconducting materials whose electronic properties are largely dependent on the number of electrons in the f-shell. Therefore, the importance of lanthanides in forming superconductors with desirable properties is exceptional in forming functional superconductors with outstanding technological importance.

2.4   Applications of LBSC

Superconductors are widely used in applications requiring high current densities, very large magnetic fields, and devices that must be energy efficient. Most of LBSC can be utilized in these applications since they possess the above said attributes. Recent work has shown that LBSC are potential superconducting materials with promising technological applications [18, 19, 23]. The technological applications of superconductors depend on the Tc as well. LBSC can either be low temperature superconductors (LTSC) or high temperature superconductors. LTSC are important components of magnetic resonance imaging (MRI) and nuclear magnetic resonance (NMR) equipment. Magnetic resonance imaging requires powerful magnets which are the major components of an MRI and NMR scanners. Most of superconducting magnets exhibit magnetic field of several Tesla and hence suitable for this application [32, 33]. Generally, the utilization of LTSC in these applications is still convenient since they are less expensive in terms of preparation and maintenance compared to the HTSC.

In transportation superconductors can also play a very vital role. Superconductors can repel the magnetic field of magnets a phenomenon called as the perfect diamagnetism [32]. As a result, superconductors are used to levitate trains to provide viable high-speed without friction and hence minimize energy expenditure.

Conclusions

The current work provided a general understanding of the importance, preparation and characterization of lanthanide-based superconductors. One key importance of lanthanides is their ability to form both low and high temperature superconductors due to their exceptional electronic properties contributed by the existence of the f orbital electrons. Indeed, their importance of being used as dopants in other classes of superconductors generates final products with unique lattice structures and electronic properties. The discovery of lanthanide superhydrides as HTSC is expected to revolutionize the formation of functional superconducting materials since their Tc can reach up to 260 K. The physico-chemical properties exhibited by lanthanide superhydrides indicate that this class of materials is very vibrant and is expected to solve most of the practical applicability encountered by HTSC. Therefore, designing cost-effective methods to synthesize LBSC with desirable properties for various applications is in great demand and commendable.

Superconductors - Materials and Applications   Materials Research Forum LLC
Materials Research Foundations 132 (2022) 97-107   https://doi.org/10.21741/9781644902110-5

References

[1] H. Hosono, A. Yamamoto, H. Hiramatsu, Y. Ma, Recent advances in iron-based superconductors toward applications, Mater. Today, 21 (2018) 278-302. https://doi.org/10.1016/j.mattod.2017.09.006

[2] D. Larbalestier, A. Gurevich, D.M. Feldmann, A. Polyanskii, High-Tc superconducting materials for electric power applications, Materials For Sustainable Energy: A Collection of Peer-Reviewed Research and Review Articles from Nature (2011) 311-320. https://doi.org/10.1142/9789814317665_0046

[3] X. Geng, J. Yi, The development of high-temperature superconductors and 2D iron-based superconductors, in: Nano-sized multifunctional materials, Elsevier (2019) 117-144. https://doi.org/10.1016/B978-0-12-813934-9.00006-2

[4] Y. Lvovsky, P. Jarvis, Superconducting systems for MRI-present solutions and new trends, IEEE Trans. Appl. Supercond. 15 (2005) 1317-1325. https://doi.org/10.1109/TASC.2005.849580

[5] M. Schmelz, R. Stolz, V. Zakosarenko, S. Anders, L. Fritzsch, H. Roth, H.G. Meyer, Highly sensitive miniature SQUID magnetometer fabricated with cross-type Josephson tunnel junctions, Physica C Supercond. 476 (2012) 77-80. https://doi.org/10.1016/j.physc.2012.02.025

[6] Y. Wu, T. van Ree, Introduction: Energy technologies and their role in our life, in: Metal Oxides in Energy Technologies, Elsevier (2018) 1-16. https://doi.org/10.1016/B978-0-12-811167-3.00001-8

[7] S. Matsuishi, T. Hanna, Y. Muraba, S.W. Kim, J.E. Kim, M. Takata, S.-i. Shamoto, R.I. Smith, H. Hosono, Structural analysis and superconductivity of $CeFeAsO1-xHx$, Phys. Rev. B. 85 (2012) 014514. https://doi.org/10.1103/PhysRevB.85.014514

[8] G. Biswal, K. Mohanta, A recent review on iron-based superconductor, Mater. Today: Proc. 35 (2021) 207-215. https://doi.org/10.1016/j.matpr.2020.04.503

[9] J.G. Bednorz, K.A. Müller, Possible high T c superconductivity in the Ba− La− Cu− O system, Phys. B: Condens Matter. 64 (1986) 189-193. https://doi.org/10.1007/BF01303701

[10] H. Hosono, K. Tanabe, E. Takayama-Muromachi, H. Kageyama, S. Yamanaka, H. Kumakura, M. Nohara, H. Hiramatsu, S. Fujitsu, Exploration of new superconductors and functional materials, and fabrication of superconducting tapes and wires of iron pnictides, Sci. Technol. Adv. Mater. (2015). https://doi.org/10.1002/chin.201551196

[11] M.R. Koblischka, S. Roth, A. Koblischka-Veneva, T. Karwoth, A. Wiederhold, X.L. Zeng, S. Fasoulas, M. Murakami, Relation between crystal structure and transition temperature of superconducting metals and alloys, Met. 10 (2020) 158. https://doi.org/10.3390/met10020158

[12] A. Drozdov, P. Kong, V. Minkov, S. Besedin, M. Kuzovnikov, S. Mozaffari, L. Balicas, F. Balakirev, D. Graf, V. Prakapenka, Superconductivity at 250 K in lanthanum hydride under high pressures, Nature. 569 (2019) 528-531. https://doi.org/10.1038/s41586-019-1201-8

[13] M.-K. Wu, J.R. Ashburn, C. Torng, P.H. Hor, R.L. Meng, L. Gao, Z.J. Huang, Y. Wang, a. Chu, Superconductivity at 93 K in a new mixed-phase Y-Ba-Cu-O compound system at ambient pressure, Phys. Rev. Lett. 58 (1987) 908. https://doi.org/10.1103/PhysRevLett.58.908

[14] A. Schilling, M. Cantoni, J. Guo, H. Ott, Superconductivity above 130 k in the Hg-Ba-Ca-Cu-O system, Nature. 363 (1993) 56-58. https://doi.org/10.1038/363056a0

[15] M. Somayazulu, M. Ahart, A.K. Mishra, Z.M. Geballe, M. Baldini, Y. Meng, V.V. Struzhkin, R.J. Hemley, Evidence for superconductivity above 260 K in lanthanum superhydride at megabar pressures, Phys. Rev. Lett. 122 (2019) 027001. https://doi.org/10.1103/PhysRevLett.122.027001

[16] P. Baker, S. Giblin, F. Pratt, R. Liu, G. Wu, X. Chen, M. Pitcher, D. Parker, S. Clarke, S. Blundell, Heat capacity measurements on FeAs-based compounds: a thermodynamic probe of electronic and magnetic states, New J. Phys.11 (2009) 025010. https://doi.org/10.1088/1367-2630/11/2/025010

[17] Y. Kamihara, T. Watanabe, M. Hirano, H. Hosono, Iron-based layered superconductor La [O1-x Fx]FeAs (x= 0.05− 0.12) with T c= 26 K, J. Am. Chem. Soc. 130 (2008) 3296-3297. https://doi.org/10.1021/ja800073m

[18] H. Song, Z. Zhang, T. Cui, C.J. Pickard, V.Z. Kresin, D. Duan, High T c Superconductivity in Heavy Rare Earth Hydrides, Chin. Phys. Lett. 38 (2021) 107401. https://doi.org/10.1088/0256-307X/38/10/107401

[19] H. Song, Z. Zhang, T. Cui, C.J. Pickard, V.Z. Kresin, D. Duan, High Tc superconductivity in heavy Rare Earth Hydrides: correlation between the presence of the f states on the Fermi surface, nesting and the value of Tc, (2020) p. arXiv:2010.12225

[20] H.R. Ott, Ten years of superconductivity: 1980-1990, Springer Science & Business Media, 7 (2012).

[21] Z.C. Wang, C.Y. He, S.-Q. Wu, Z.T. Tang, Y. Liu, A. Ablimit, Q. Tao, C.M. Feng, Z.A. Xu, G.-H. Cao, Superconductivity at 35 K by self doping in RbGd2Fe4As4O2, J. Phys. Condens. Matter. 29 (2017) 11LT01. https://doi.org/10.1088/1361-648X/aa58d2

[22] S.Q. Wu, Z.C. Wang, C.Y. He, Z.T. Tang, Y. Liu, G.-H. Cao, Superconductivity at 33-37 K in A L n 2 Fe 4 As 4 O 2 (A= K and Cs; L n= lanthanides), Phys. Rev. Mater. 1 (2017) 044804.

[23] R. Hott, R. Kleiner, T. Wolf, G. Zwicknagl, Superconducting materials-A topical overview, Fron. Supercond. Mater. (2005) 1-69. https://doi.org/10.1007/3-540-27294-1_1

[24] Z.M. Geballe, H. Liu, A.K. Mishra, M. Ahart, M. Somayazulu, Y. Meng, M. Baldini, R.J. Hemley, Synthesis and stability of lanthanum superhydrides, Angew. Chem. 130 (2018) 696-700. https://doi.org/10.1002/ange.201709970

[25] H. Liu, I.I. Naumov, Z.M. Geballe, M. Somayazulu, S.T. John, R.J. Hemley, Dynamics and superconductivity in compressed lanthanum superhydride, Phys. Rev. B. 98 (2018) 100102. https://doi.org/10.1103/PhysRevB.98.100102

[26] Y. Qi, J. Guo, H. Lei, Z. Xiao, T. Kamiya, H. Hosono, Superconductivity in noncentrosymmetric ternary equiatomic pnictides LaMP (M= Ir and Rh; P= P and As), Phys. Rev. B. 89 (2014) 024517. https://doi.org/10.1103/PhysRevB.89.024517

[27] J. Guo, S. Jin, G. Wang, S. Wang, K. Zhu, T. Zhou, M. He, X. Chen, Superconductivity in the iron selenide K x Fe2Se2 ($0 \leq x \leq 1.0$), Phys. Rev. B. 82 (2010) 180520. https://doi.org/10.1103/PhysRevB.82.180520

[28] Y. Guo, X. Wang, J. Li, Y. Sun, Y. Tsujimoto, A.A. Belik, Y. Matsushita, K. Yamaura, E. Takayama-Muromachi, Continuous critical temperature enhancement with gradual hydrogen doping in LaFeAsO 0.85 H x (x= 0-0.85), Phys. Rev. B. 86 (2012) 054523. https://doi.org/10.1103/PhysRevB.86.054523

[29] S. Iimura, S. Matsuishi, H. Sato, T. Hanna, Y. Muraba, S.W. Kim, J.E. Kim, M. Takata, H. Hosono, Two-dome structure in electron-doped iron arsenide superconductors, Nat. Commun. 3 (2012) 1-7. https://doi.org/10.1038/ncomms1913

[30] D. Zhou, D.V. Semenok, D. Duan, H. Xie, W. Chen, X. Huang, X. Li, B. Liu, A.R. Oganov, T. Cui, Superconducting praseodymium superhydrides, Sci. Adv. 6 (2020) eaax6849. https://doi.org/10.1126/sciadv.aax6849

[31] A.M. Shipley, M.J. Hutcheon, M.S. Johnson, R.J. Needs, C.J. Pickard, Stability and superconductivity of lanthanum and yttrium decahydrides, Phys. Rev. B. 101 (2020) 224511. https://doi.org/10.1103/PhysRevB.101.224511

[32] C.J. Kim, Applications of Superconductors, in: Superconductor Levitation, Springer, (2019) 213-236. https://doi.org/10.1007/978-981-13-6768-7_11

[33] P. Komarek, Advances in large scale applications of superconductors, Supercond. Sci. Technol. 13 (2000) 456. https://doi.org/10.1088/0953-2048/13/5/304

[34] T. Hanna, Y. Muraba, S. Matsuishi, N. Igawa, K. Kodama, S. Shamoto, H. Hosono, Hydrogen in layered iron arsenides: Indirect electron doping to induce superconductivity, Phys. Rev. B. 84 (2011) 024521. https://doi.org/10.1103/PhysRevB.84.024521

Superconductors - Materials and Applications

Materials Research Foundations 132 (2022) 108-130

Materials Research Forum LLC

https://doi.org/10.21741/9781644902110-6

Chapter 6

Type I Superconductors: Materials and Applications

Prashant Hitaishi[1], Rohit Verma[2*], Parul Khurana[3] and Sheenam Thatai[2]

[1]Department of Physics, School of Natural Sciences, Shiv Nadar University, G.B. Nagar, 201314, India

[2]Amity Institute of Applied Sciences, Amity University, Noida, 201313, India

[3]G.N.Khalsa College, Mumbai University, Mumbai, 400019, India

* rverma85@amity.edu

Abstract

Superconductors are materials that show properties of perfect diamagnetism and zero resistivity or infinite conductivity below a well-defined temperature known as critical temperature and, in the presence of a magnetic field that is less than a critical value. Based on their behaviour to the externally applied magnetic field, superconductors are divided into two categories; Type-I and Type-II superconductors. In this chapter, we have described various Type-I superconductors, their critical temperatures, history of superconductivity, physical properties, applications, and recent researches on Type-I superconductors, various theories proposed for explaining their properties, and different applications of superconductors.

Keywords

Superconductors (SCs), Type-I SCs, Meissner Effect, Theories of Low-Temperature Superconductors, Applications of Superconductors

Contents

1. Introduction

The word "superconductivity" comprises of two words 'super' and 'conductivity' that means extremely high conductivity. The first thought that comes to our mind with the word superconductivity is a phenomenon or property of specific material to lose its resistance at low temperature or simply the phenomenon of vanishing the resistivity below a critical temperature. But it is not the only remarkable characteristic of superconductors.

Heike Kamerlingh Onnes is known as the father of superconductivity. He was the first to liquefy helium in 1908, used it to attain a low temperature, and successfully measured the resistivity of various metals. In 1911, he was working in his Leiden laboratory and observed the strange properties of mercury at low temperatures. Initially, he called it *suprageleider* (in Dutch), translated into English, *supraconductivity* but became famous as *superconductivity* [1-3].

The discovery of Onnes' was something different from the material's properties known at that time. Metals conduct electricity due to free electrons, while in insulators, electrons are bound and not free to move. Semiconductor lies between these two extremes, metals and insulators but behaves more like an insulator. Hence impurities are added to enhance the conductivity of semiconductors. It was observed that the electrical resistance of mercury vanished below a specific critical temperature (T_C = 4.2 K) and exhibit zero resistivity. The abrupt fall in the resistance at a critical temperature T_C is referred to as the phase transition of a material from a normal to a superconducting state. It is not a low resistance state but a true zero resistance. Not all metals become superconductors at low temperatures, *so what is necessary for a metal to be a superconductor?*

Superconductors are broadly classified into two categories: Type-I and Type-II superconductors. In this chapter, we will discuss the history of superconductivity, physical properties, applications, and recent researches on Type-I superconductors.

2.   Type-I superconductors

Initially, thirty pure metals were listed as Type-I superconductors but later three of them were identified as type-II hence removed from that list. All thirty Type-I superconductors with their physical properties like critical temperature and magnetic field are tabulated in table 1. Last three elements Vanadium (V), Technetium (Tc), and Niobium (Nb) are shown in blue color because they behave like type-II superconductors. That's why removed from the list of Type-I superconductors. Type-I superconductors exhibit properties like zero resistivity below T_C, low critical magnetic field, zero internal magnetic fields, follow BCS theory, and loses superconductivity easily at critical externally applied magnetic field hence known as *"soft superconductors"*. Metals like gold (Au), silver (Ag), and copper (Cu) are known for their superior electrical conductivity at room temperature but do not show superconductivity in any situation.

In the superconducting state, pairs of an electron having anti-aligned spin and momenta interact with the lattice and hence develop an extremely weak magnetic attraction for each other [4]. Superconductivity depends on the interaction between atoms in the lattice and electron pairs. Coupling between electron and phonon plays an important role in the zero resistivity of superconductors. This coupling is very little in good conductors because electrons get scattered by phonons and increase the resistance [5]. At room temperature, best conductors like silver, copper, and gold have the smallest lattice vibrations hence do not become a superconductor. Superconductivity only exists below the critical temperature and magnetic field strength [4].

Table 1: List of thirty type-I superconductors with their corresponding critical temperature and magnetic field.

S.No.	Material	Critical temperature [Tc] (in Kelvin)	Critical magnetic field Ho [in mT]	Reference
1	Rhodium (Rh)	3.25×10^{-4}	4.9×10^{-3}	[1,6]
2	Tungsten (W)	0.011, 0.0154	0.115	[1,5,7,8]
3	Beryllium (Be)	0.026	6.8	[1,9]
4	Iridium (Ir)	0.1125-0.14	1.6	[1,2,5,8]
5	Lutetium (Lu)	0.1	35	[1,4,5,8]
6	Hafnium (Hf)	0.1-0.165	1.27	[1,2,4,5,7,8]
7	Ruthenium (Ru)	0.5	6.9	[4,5,8]
8	Osmium (Os)	0.7	7	[1,4,5,8]
9	Molybdenum (Mo)	0.915	9.6	[1,5,8]
10	Zirconium (Zr)	0.5-0.61	4.7	[1,4,5,8]
11	Cadmium (Cd)	0.5-0.52	2.8	[2,4,5,8,10]
12	Uranium (U)	0.2, 0.7, 1.8	-	[1,4,5,7,8]
13	Titanium (Ti)	0.39-0.5	5.6	[1,2,4,5,8]
14	Zinc (Zn)	0.85-0.855	5.4	[1,5,8,11]
15	Gallium (Ga)	1.08-1.1	5.83	[1,4,8,11]
16	Gadolinium (Gd)			
17	Aluminium (Al)	1.18-1.2	10.49	[1,4,8,12]
18	Protactinium (Pa)	1.4	-	[7,8]
19	Thorium (Th)	1.38, 1.4	0.16	[1,5,7,8]
20	Rhenium (Re)	1.697-1.7	20	[1,5,8]
21	Thallium (Tl)	2.38-2.4	17.8	[1,4,5,8]
22	Indium (In)	3.408	28.15	[1,2,4,8,10]
23	Tin (Sn)	3.72	30.5	[1,4,5,8]
24	Mercury (Hg)	3.95-4.15	41.1, 33.9	[1,2,8,10]
25	Tantalum (Ta)	4.4-4.5	82.9	[1,4,5,7,8]
26	Lanthanum (La)	4.88, 6	80, 109.6	[1,4,5,8]
27	Lead (Pb)	7.196	80.3	[4,5,8]
28	Vanadium (V)	5.4	140.8	[1,4,5,8]
29	Technetium (Tc)	7.8	141	[4,5,8]
30	Niobium (Nb)	9.25	206	[1,5,8]

3. History of superconductivity

3.1. Quest for low temperature

The discovery of superconductivity until 1911 is divided into various phases that include interesting research and essential developments in science. The main challenges were liquefaction and condensation of gases to attain low temperatures. The accidental discovery of the liquefaction process of chlorine played a vital role in the discovery of superconductors. Carl Scheele discovered chlorine in 1774, and after intensive work for decades Davy and Michael Faraday finally made the liquid chlorine in 1823. Faraday's experiments developed an understanding that a high pressure created using a sealed tube and a low temperature of -34 ̊C were required to liquefy the chlorine gas. An increase in the pressure helped to raise the boiling point of chlorine. Faraday could not liquefy hydrogen, oxygen and nitrogen gas because the high pressure available in Faraday's lab was insufficient to raise their boiling temperature. He called them permanent gases; later in 1860, Thomas Andrews worked on the gas laws and conditions like pressure, temperature, and volume required to liquefy a gas. In 1852, William Thomson and James Prescott Joule established the cooling principle by rapid expansion, which is known as Joule-Kelvin effect [1,13].

3.2 Discovery of Helium

In 1868, Pierre Janssen noticed a yellow line in the sunlight during a total solar eclipse and estimated its wavelength to be 587.49 nm. Initially, sodium was assumed to cause that yellow line but later, Norman Lockyer and Edward Frankland confirmed the origin of that line due to a new helium element. In 1895, William Ramsay successfully isolated helium from mineral cleveite. In 1904, he was awarded the Nobel Prize to discover the inert gaseous elements like helium, argon, neon, krypton, and xenon in air and their places in the periodic system. After discovering helium gas, Sir James Dewar was the first person in the race to liquefy hydrogen and helium gas. Finally, Dewar announced that he had obtained hydrogen as a static liquid after working for years in 1898. Sir James Dewar had expertise in low temperature, and Ramsay had an excellent knowledge of helium. They could be the first to liquefy helium if they work together but unfortunately failed to do so.

On the other side, Van der Waals was working to understand the properties of gases and realized the discrepancy in the ideal gas theory. Robert Boyle completely ignored the intermolecular forces between gas molecules in the theory of ideal gas; hence liquefaction of gases on cooling could not be predicted. In 1880, Van der Waals considered the intermolecular forces and published his real gas equation applicable to all real gases. That equation successfully established the relation between temperature and liquefaction strength of a gas. In 1881, Onnes set up his world-famous low-temperature physics laboratory in Leiden. He successfully liquefied hydrogen eight years after Dewar in 1906, but his apparatus was more reliable and production quantity was more than Dewar. Finally, in 1908 he made the liquid helium in the laboratory and measured its temperature as 4.2 K. In 1913, Kamerlingh Onnes was awarded "The Nobel Prize for Physics" to investigate the properties of matter at low temperatures. He acknowledged the van der Waals for his theory and support that helped him in the liquefaction of gases. Onnes' lab in Leiden had a monopoly on low-temperature superconductivity work for a long time because they had a supply of liquid helium in sufficiently large quantity [1]. Different gases and compounds used as a coolant to study the properties of conductors at low temperature is shown in Fig. 1.

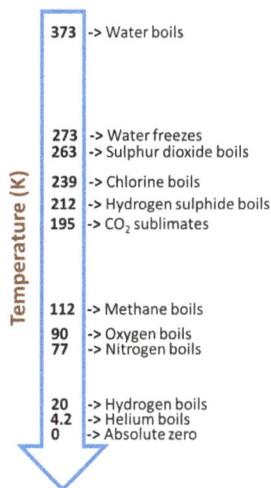

373	-> Water boils
273	-> Water freezes
263	-> Sulphur dioxide boils
239	-> Chlorine boils
212	-> Hydrogen sulphide boils
195	-> CO_2 sublimates
112	-> Methane boils
90	-> Oxygen boils
77	-> Nitrogen boils
20	-> Hydrogen boils
4.2	-> Helium boils
0	-> Absolute zero

Figure 1.   Boiling point of various compounds and gases. Liquid helium and nitrogen boil at 4.2K and 77K respectively.

3.3   Curiosity to know the resistance of metals at absolute zero?

After the liquefaction of helium, Onnes was curious to investigate the properties of metals at low temperatures and to know the resistance of metal at absolute zero. It was well known that electrons flow through a solid on an application of electric current across it. Electrons interact with the vibrating atoms and get deflected from the path known as the scattering of electrons. At low temperatures, the atomic vibrations are freeze means less scattering, less deflection hence low resistance. At the beginning of the 20th century, there were three possible theories to define the resistance of metals at low temperature. According to the theories of Lord Kelvin, Matthiessen, and Dewar, resistance of metals was shown in Fig. 2. Dewar worked on silver and gold and was able to cool them up to 16K, but these metals did not follow the trend valid for a superconductor. Onnes noticed the resistance of platinum and gold decreases with the decrease in temperature and become constant but, non-zero. He concluded that the presence of slight impurities could not be neglected even in pure gold [3].

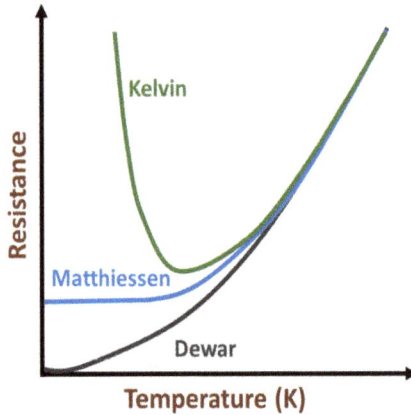

Figure 2. *Three famous theories at the beginning of 20th century for the resistance of metals at low temperature.*

3.4 Why mercury used to measure low-temperature resistance?

Impurities in the metals were always a problem in the resistance measurement as they act as a scattering center for free electrons and hinder their motion. That was one of the main reasons for Onnes to focus especially on liquid metal. He could purify the liquid metal mercury by repeated distillation process and fill it in U-shape capillaries to make mercury wires. Onnes had an advantage of his newly discovered liquid helium that helped him achieve lower temperatures than Dewar [1]. Finally, in 1911, experiments were performed to measure the resistance of mercury at a temperature below the boiling point of helium, i.e. 4.25 K [3]. It was noticed that the resistance of mercury drops to millionth part at a specific temperature. Repeated experiments confirmed the sudden fall in the resistance at a specific critical temperature is the actual behaviour of mercury rather than any experimental error. Onnes also noticed that the transition behaviour for mercury was consistent even after adding few impurities in it. Such bizarre transition was not limited to mercury only. In 1912, tin and lead also showed such behaviour at a critical temperature of 3.8K and 7.19K, respectively [1,3,4]. And this behaviour was named "superconductivity." A comparison of resistance between mercury, platinum and gold is shown in Fig. 3. This figure is traced from the graph published in "K. Onnes, Commun. Phys. Lab. 12, 120 (1911)" and reference [3].

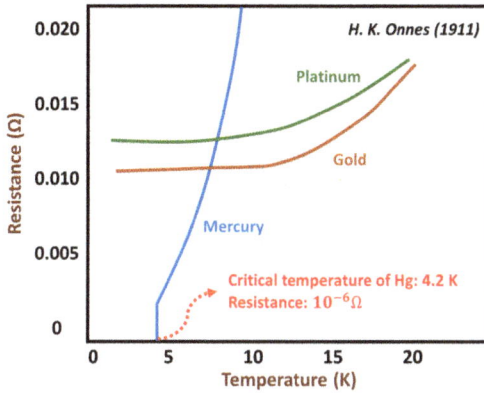

Figure 3.   The electrical resistance of platinum, gold and mercury at low temperature traced from data plotted by H.K. Onnes in 1911.

4.   Attributes of superconductors

4.1   Current in a superconductor coil

In the absence of an external source, the power loss (P) is directly proportional to the resistance (R) of the conductor and the amount of current (I) allowed to pass through it (P $= I^2 R$). In an ordinary conductor, current decays rapidly due to the energy dissipated in the form of heat, while it is not the case in a superconductor, and current sustains without attenuation due to its zero resistance. A current-carrying wire gets hot due to a well-known Joule heating phenomenon is absent in superconductors. A superconducting wire coil carrying an electrical current would sustain current forever even in the absence of an external power source and keep going around and around the coil [1]. Onnes successfully constructed a circuit to observe the continuous flow of current over a long period, even after removing any active power source. A battery was connected initially to allow the flow of current in the circuit and then disconnected, but the current continued to flow. He demonstrated zero resistance of superconductors by measuring the decay of current in superconductor rings. The sustainability of induced current has been studied in various metal rings to demonstrate the zero resistance in superconductors. It was observed that induced current decayed rapidly in an ordinary metal ring while it sustains for years in lead rings (T_C = 7.2 K) [4].

4.2   How superconductors behave in an external magnetic field?

The zero resistance or superconducting state achieved at a low temperature could be destroyed if the material was placed in a sufficiently large external magnetic field or current exceeded a critical value. Onnes was interested in using superconductor wires to fabricate large magnetic coils. But experiments showed that critical magnetic field and current are two sides of a coin, and superconductivity is destroyed even if the applied magnetic field is small. There were two main challenges to use superconductors for potential applications. One was to discover a superconductor with a high transition temperature and another a high critical magnetic field. It took decades to solve these two issues; as a result, various new superconductors were discovered.

According to Michael Faraday's law of magnetic induction, a superconducting wire carries an infinite amount of current; hence the magnetic field induced in it could never be changed. It was thought that the magnetic field induced initially would remain constant and trapped in the superconductor unless its temperature was raised above the transition temperature. Experimental results showed such a type of field trapping in the superconductors.

In 1931, W. J. de Haas and W. H. Keesom at Leiden laboratory discovered an alloy superconductor. Interestingly elements used in the alloy did not show superconductivity, but alloy showed an enhanced critical magnetic field. It was a great success in this field with new hope and direction for research.

In 1933, Walther Meissner and Robert Ochsenfeld performed experiments on superconductors at low temperatures to study the effect of externally applied magnetic field around the superconductor. They observed that the magnetic field was expelled around the superconductor rather than trapping, contrary to the earlier experiments. Meissner suggested the trapping behaviour observed earlier was only due to the presence of impurities in the sample. A superconductor electrical current did not pass through it, only flowing across its surface and screen the interior from an externally applied magnetic field, hence known as *screening currents*. The expulsion of an external magnetic field is due to the presence of a magnetic field produced by screening currents that act opposite to the external field. This phenomenon of forcible expulsion of the magnetic field in the superconductors is known as the *Meissner effect.*

4.3   Unification of electric and magnetic behaviour

The well-known Ohm's law states the relation between applied voltages, current passed through metal and its electrical resistance. It was based on classical theory and needed to be updated according to the quantum theory. Paul Drude gave an electron gas model and generalized Ohm's law. Drude's model was also a classical model but Sommerfeld modified it using Fermi-Dirac statistics in 1927, known as the Drude-Sommerfeld model. This model could be applicable for ordinary metals, not for superconductors.

The London brothers were interested in working for superconductors and trying to find generalized relations like Ohm's law and Drude-Sommerfeld model. They considered the Meissner effect as a fundamental characteristic of superconductors and noticed that electrons in the superconducting state behave differently than in a normal state. According to them, electrons in a conductor are independent of each other, but in a superconductor, electrons are interconnected and behave as a single entity. In 1935, finally, the London brothers wrote equations for superconductors that relate the screening current flowing across the surface to magnetic field. London theory suggested an exponential decay of the magnetic field rather than a sudden drop. Field could penetrate the superconductor's surface to a certain distance known as the *characteristic length or London penetration depth*. This is the distance at which the value of current falls to exponential (1/e) of current on the surface of superconductor. In this theory, electrons were divided into two types as normal and super electrons present in normal and superconducting state respectively. Below transition temperature only super electrons are present and the number of normal electrons increases with the increase in temperature. Finally, above transition temperature only normal electrons are there. London equations successfully explained the Meissner effect and zero resistivity but failed to explain exactly what happened inside the superconductor [1].

5. Characteristics of type-I superconductors

5.1 Critical Temperature (T_C)

Phase transition from normal to superconductor and abrupt fall in resistance occurred at a specific temperature known as *critical temperature* (T_C). Type-I superconductors have very low transition temperature hence, maintaining that temperature is not that feasible. Mercury has a transition temperature of 4.2 K that can be achieved using liquid helium. In 1911, Onnes measured the resistance of gold, platinum and mercury at low temperature as shown in Fig. 3. But using liquid helium for practical applications is not economical. Each element has its critical temperature. The electrical and magnetic properties of superconductors changes with the temperature and material switches from one state to another. This change from normal to superconducting state or vice-versa is defined as a phase transition. Type-I superconductors are arranged in the increasing order of critical temperature in table 1.

5.2 Meissner effect or perfect diamagnetism

Superconductors discovered by the Onnes act as a perfect conductors and show perfect diamagnetic behaviour in the superconducting state. Screening currents are induced on the surface of superconductors when placed in external varying magnetic field. It means they tend to repel the magnetic field lines, hence vanishing the field inside it. Meissner effect is reversible and the perfect diamagnetic behaviour is independent of the superconductor's history as shown in Fig. 4.

If a superconductor is placed in an externally applied magnetic field intensity H then, magnetic flux density B inside it is zero ($\vec{B} = 0$) according to Meissner effect. We can write,

$$\vec{B} = \mu_0(\vec{H} + \vec{M}) \tag{1}$$

$$0 = \mu_0(\vec{H} + \vec{M}) \tag{2}$$

$$\vec{H} = -\vec{M} \tag{3}$$

$$\chi = \frac{\vec{M}}{\vec{H}} = -1 \tag{4}$$

Hers, M is magnetization and χ is magnetic susceptibility. For perfectly diamagnetic materials, magnetic susceptibility χ is -1, which is valid for type-I superconductors. The expulsion of magnetic field is observed up to a particular value of externally applied magnetic field. That is known as the *critical magnetic field* for a specific superconductor. As shown in Fig. 5 beyond this critical field type-I superconductor shows an abrupt transition from superconducting to normal state. The superconductivity of type-I materials vanished even at low strength of field, that's why referred as "*soft superconductors*".

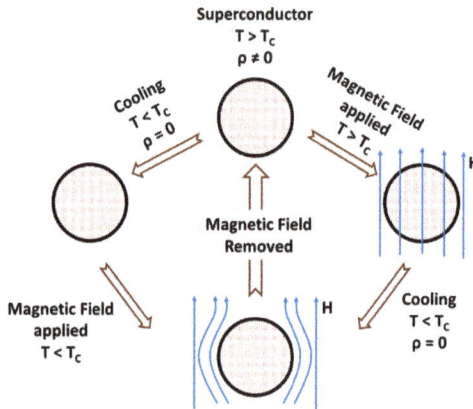

Figure 4. *Schematic representation of perfect diamagnetic behaviour in a type-I superconductor is independent of its past history.*

A zero magnetic field (perfectly diamagnetic behaviour) is independent of the zero resistance property and does not necessarily rely on it. According to Ohm's law ($\vec{E} = \rho\vec{J}$), electric field cannot exist inside a perfect conductor ($\rho = 0$) means $\vec{E} = 0$. According to Maxwell's equation:

$$\left(\nabla \times \vec{E} = -\frac{\partial \vec{B}}{\partial t} \right) \tag{5}$$

$$\vec{B} = constant \tag{6}$$

It means a perfect conductor would trap magnetic field in it as shown in Fig. 6. Perfect diamagnetism means $\vec{B} = 0$, magnetic flux density inside material is zero. Superconductivity is defined by two independent properties, $\vec{B} = 0$ and $\vec{E} = 0$.

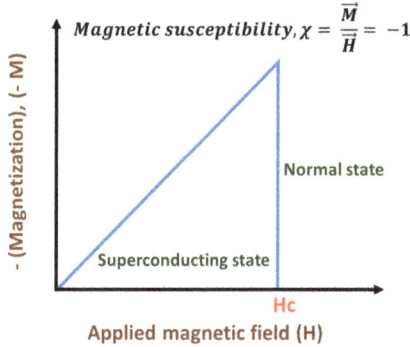

Figure 5.   *Magnetization (M) response of type-I superconductors according to the externally applied magnetic field (H).*

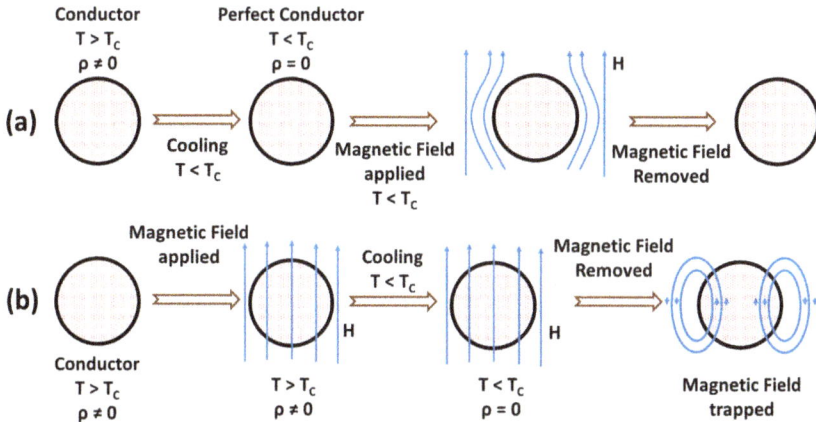

Figure 6.   *Schematic representation of conductors in different cases, (a) A perfect conductor first cooled then placed in an external magnetic field. (b) A conductor first placed in a magnetic field then cooled to achieve perfect conductor state.*

5.3 Critical magnetic field (H$_C$)

Type-I superconductor vanishes their superconductivity in the presence of sufficiently large magnetic field. The maximum externally applied field in which a superconductor can withstand without losing its superconductivity is known as critical magnetic field (H$_C$). It is a material and temperature dependent property. It varies with the temperature and follows equation 7. Here, H$_C$ (T) is the critical magnetic field at a T Kelvin and H$_0$ is the magnetic field at an absolute zero temperature [4,12]. It shows a parabolic relation between temperature and applied magnetic field. Plot in Fig. 7 is known as the magnetic phase diagram as it denotes the boundary between superconducting and normal state. It is interesting to note from table 1 that elements having high critical temperature exhibit relatively higher critical magnetic field. Type-I superconductors exhibit a stable superconducting state only for a definite range of temperature and magnetic field. If we keep the sample beyond this defined range, the normal state become more stable than superconducting.

$$H_C(T) = H_0 \left[1 - \left(\frac{T}{T_C} \right)^2 \right] \tag{7}$$

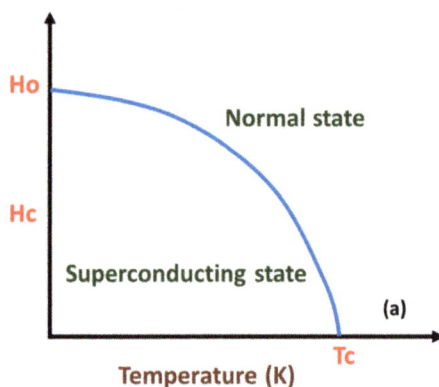

Figure 7.   Magnetic phase diagram for type-I superconductor. Relation between applied temperature and magnetic field

5.4 Critical current (I$_C$)

According to Ampere's law, a long current (I) carrying wire produces a magnetic field (B) around it. If this induced field is greater than the critical magnetic field of that superconductor then its superconductivity will be destroyed. Type-I superconductors have very low critical magnetic field values. To maintain its superconductivity a limited amount of current can be passed through the wire. That maximum applicable current is known as *critical current* for a particular material.

If a superconductor wire having a radius R and critical magnetic field *(Hc)* then its critical current *(Ic)* can be calculated using Ampere's law,

$$I_C = \frac{H_C \times (2\pi R)}{\mu_0} \tag{8}$$

As shown in Fig. 8 when an external electric field is applied on a conductor, charges in it arranged themselves in such a manner that net electric field inside conductor is zero. External electric field lines do not penetrate a normal conductor. If current is applied on a superconductor and it is placed in an external magnetic field ($H_{external}$) the current flow across its surface. This current is known as surface or supercurrent. A magnetization (M) is produced in the superconductor due to the flow of supercurrents that cancels the $H_{external}$. Net flux density, $\vec{B}$ inside a superconductor is zero as a result magnetic field lines do not penetrate a superconductor [4].

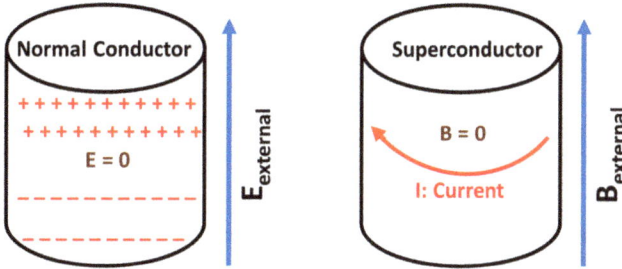

Figure 8.   Schematic representation of behaviour of a normal conductor and superconductor in the external electric and magnetic field respectively.

5.5   Isotope effect

The atoms in a crystalline solid are arranged in a periodic arrangement known as crystal lattice. And are mostly found moving and vibrating randomly, but at absolute zero atoms sit perfectly. A theoretician, Herbert Fröhlich was interested to understand the effect of atomic vibrations in crystal on superconductivity. The motion of atoms depends on the temperature of solid and vibration frequency is inversely related to the square root of atomic mass. If the frequency of atomic vibrations is related to superconductivity then there could be a relation between transition temperature and atomic mass. Emanuel Maxwell and Bernard Serin were the first who experimentally studied the effect of atomic mass on transition temperatures [1]. Superconductivity involves both electronic effects and lattice vibrations. In 1950, the isotope effect was demonstrated, that confirms an important relation between atomic mass, vibration, and superconductivity. Transition temperature (T_C) of mercury isotopes was studied by two different groups as shown in Fig. 9. It followed an inverse relation with the atomic mass (A) of isotope $\left(T_C \propto \frac{1}{\sqrt{A}}\right)$. This graph is plotted using data mentioned in following references [14,15]. Dash line is the linear fit to show the consistency of results obtained by different groups.

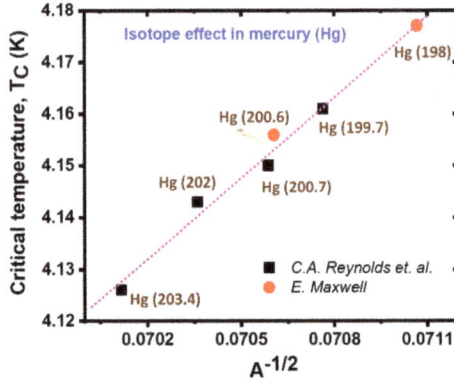

Figure 9.   *Isotope effect in mercury: critical temperature (T_C) depends on the atomic mass of Hg isotope. Superconductivity also involves the lattice vibrations, it is not purely an electronic effect.*

5.6   Development of theories of superconductivity

Theoretical research played a very crucial role in the development of superconductivity. They gave us microscopic insight to various macroscopic phenomena observed during experiments. Both experimental and theoretical models have been modified to have a better understanding. In this section we will focus on three main theories: London and London theory (1935), Ginzburg and Landau theory (1950), and BCS theory (1957).

5.6.1   London equations and penetration depth

In 1935, London brothers (Fritz and Heinz London) gave a phenomenological theory of superconductivity that satisfied both experimentally observed phenomena, Meissner effect and infinite conductivity. According to London theory, two type of electrons named normal and super electrons are present in a superconductor. Below critical temperature all are super electrons but with the increase in temperature, super electrons decrease, and normal electrons increases. Both type of electrons are considered to calculate the net current density. London states that the time derivative of current density ($\vec{J}$) is directly proportional to the electric field($\vec{E}$) [4,16]. London suggested that magnetic field strength and current density does not vanish or drop abruptly at the surface of a superconductor, it decreases exponentially as shown in the Fig. 10. A brief derivation of London's first and second equation are tabulated in Table.2 with their physical significance to superconductivity.

Table 2: Brief derivation of London's first and second equation with their physical significance.

First London equation	Second London equation
According to the London equation: $$\frac{\partial \vec{J}}{\partial t} = a \, \vec{E} \ \ ....(9)$$ Here, a is a constant of proportionality According to Ohm's law: $$\vec{J} = \sigma \vec{E} = ne\vec{V}_d \ \ .....(10)$$ Here, n is electron density and $\vec{V}_d$ is the drift velocity. **Solution for super electrons:** Equating equation 10 in 9 and taking time derivative we get, $$\frac{\partial \vec{J}_s}{\partial t} = \left(\frac{n_s e^2}{m}\right)\vec{E} \ \ ......(11)$$ $\left(\frac{n_s e^2}{m}\right)$ is the constant "a" in eq. 9 Equation 11, is known as the *first London equation.* If $\vec{E} = 0$, means $\frac{\partial \vec{J}_s}{\partial t} = 0$ $$\vec{J}_s = constant$$ • Means steady current possible even in the absence of electric field. • Describes the infinite conductivity or absence of resistivity in a superconductor. **Solution for normal electrons:** Normal electrons follow Ohm's law: $$\vec{J}_n = \sigma_n \vec{E}$$ $$\vec{J}_n = \left(\frac{n_n e^2 \tau}{m}\right)\vec{E} \ \ .....(12)$$ Equation 4 shows that $\vec{J}_n = 0$ if, $\vec{E} = 0$ • Electric field is required to flow a steady current in a normal conductor.	Starting with Maxwell's equation: $$\vec{\nabla} \times \vec{E} = -\frac{\partial \vec{B}}{\partial t} = -\mu_0 \frac{\partial \vec{H}}{\partial t} \ ...(13)$$ Taking curl of first London equation (eq 11) $$\frac{\partial (\vec{\nabla} \times \vec{J}_s)}{\partial t} = \left(\frac{n_s e^2}{m}\right)(\vec{\nabla} \times \vec{E}) \ \ .....(14)$$ Equating value of curl $\vec{E}$ in eq. 14 from eq. 13 After solution we get, $$\vec{\nabla} \times \vec{J}_s = -\frac{n_s e^2 \mu_0 \vec{H}}{m} \ \ ....(15)$$ Equation 15, is the *second London equation.* If $\vec{H} = 0$, $\vec{J}_s = constant$ • It explains the Meissner effect, zero magnetic fields inside a superconductor. **Penetration depth calculation (λ)** Taking curl of equation 15, we get $$\vec{\nabla} \times \vec{\nabla} \times \vec{J}_s = -\frac{n_s e^2 \mu_0}{m}(\vec{\nabla} \times \vec{H})....(16)$$ $$\Rightarrow \nabla^2 \vec{J}_s = \frac{1}{\lambda^2}(\vec{\nabla} \times \vec{H}) \ \ ....(17)$$ $$\lambda = \sqrt[2]{\frac{m}{n_s e^2 \mu_0}} \ \ ....(18)$$ λ is known as *penetration depth* and has a dimension of length. $$\Rightarrow \nabla^2 \vec{J}_s = \frac{\vec{J}_s}{\lambda^2} \ \ ....(19)$$ Solution of equation 19 gives, $$\vec{J}_s = \vec{J}_0 \left(e^{\frac{-x}{\lambda}}\right) \ \ .....(20)$$ Similarly, we can solve equation 17 for externally applied magnetic field H_e and get, $$\vec{H}_e = \vec{H}_0 \left(e^{\frac{-x}{\lambda}}\right) \ \ .....(21)$$

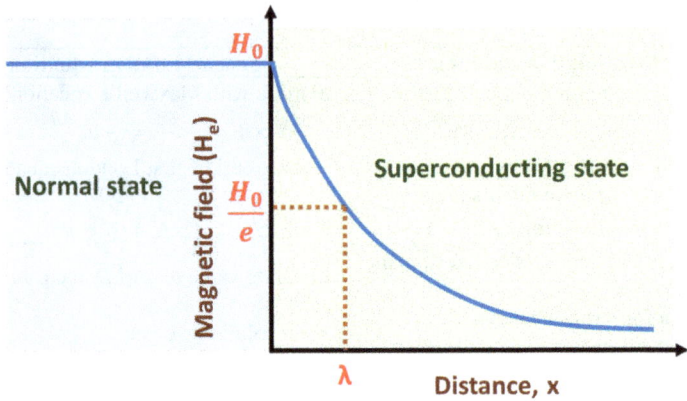

Figure 10.   Schematic plot of penetration and exponential fall of external field in the superconductor.

Now, there are two extreme cases for equation 20 and 21

At $x = 0$, $\vec{J}_s = \vec{J}_0$ and $\vec{H}_e = \vec{H}_0$

Both the current density and magnetic field are maximum and decays exponentially as we move inside superconductor from outside.

At $x = \lambda$, $\vec{J}_s = \frac{\vec{J}_0}{e}$ and $\vec{H}_e = \frac{\vec{H}_0}{e}$

Value falls to $\frac{1}{e}$ of its value at the surface.

λ is known as the *penetration depth* and has a dimension of length. It is the distance, measured from surface at which the current and field value falls to $\frac{1}{e}$ of its value at surface. London equations successfully explained the Meissner effect and zero resistivity but failed to explain exactly what happened inside the superconductor.

5.6.2 Ginzburg and Landau theory

In 1950, Vitaly Ginzburg and Lev Landau pointed out two issues in the London theory. First was that surface tension at the boundary of phase separation was not considered in it. And secondly, the destruction of superconductivity due to the current caused by super-electrons was not defined [17]. The number of super-electrons (n_s) depends on the temperature as discussed in London theory. The transition of normal and superconducting

Superconductors - Materials and Applications   Materials Research Forum LLC
Materials Research Foundations 132 (2022) 108-130   https://doi.org/10.21741/9781644902110-6

state at critical temperature (T$_C$) follows second-order transition. In this theory a new parameter is introduced to define the superconductivity known as the *order parameter (ψ)*. This parameter defined the state of order like superconducting state as an ordered state and the corresponding value of ψ equals zero (ψ = 0). While for the disordered or normal state (ψ ≠ 0). This order parameter behaves like a wave function of super-electrons (n$_s$) and is related to density of n$_s$ [4].

$$n_s = |\psi|^2 \tag{22}$$

This equation is known as the Ginzburg-Landau equation and satisfies the wave function satisfies Schrödinger like equation. Solution of the above one-dimensional second-order differential equation for wave function (ψ) :

$$\psi = Ae^{-\frac{x}{\xi}} \text{ and } \xi = \frac{\hbar}{\sqrt{2ma}} \tag{23}$$

Here, A is a constant and ξ is defined as a new parameter known as coherence length. It is the distance up to which super-electrons are correlated in terms of spins and momenta. If a superconductor is not pure and have some impurities then net coherence length (ξ_{net}) is different from inherence coherence length (ξ_s) of super-electrons. Impurities acts as the scattering center for electrons and coherence length for normal electrons is equal to the mean free path (d).

$$\frac{1}{\xi_{net}} = \frac{1}{\xi_s} + \frac{1}{d} \tag{24}$$

Table 3: Relation between coherence length and penetration depth for type-I and II superconductors.

Type-I superconductor	Type-II superconductor
• Coherence length > Penetration depth	• Coherence length < Penetration depth
• $\xi > \sqrt{2}\lambda$	• $\xi < \sqrt{2}\lambda$

The ratio of penetration depth and coherence length is defined as Ginzburg-Landau parameter (κ). The main outcome of Ginzburg-Landau equation is that it defined superconductivity in terms of quantum mechanics.

5.6.3 BCS theory

In 1957, John Bardeen, Leon Cooper and J. Robert Schieffer gave quantitative understanding and microscopic insight of superconductivity. The central theme of this theory lies in electron-lattice interaction, cooper pair formation and existence of energy gap. Electrons are fermions and occupy a discrete quantum state hence, exist electrostatic repulsion between them. But BCS theory is based on the attractive interaction of electrons caused due to virtual exchange of phonons. One electron interact with the lattice and deform in such a way that another incoming electron feel an additional force caused due to deformed lattice. As a result there is an attractive electron-electron interaction mediated by lattice vibration (phonon). Such pair of electrons present in the same quantum state shows a coherent behaviour and act like a boson. It is possible if phonon energy is greater than the difference between the energy of two interacting electrons [4,18]. Superconducting phase is possible if the interactive interaction of electrons is dominated over coulomb repulsion. Two electrons bounded together in a pair and have a size of a few hundred nanometres times greater than an atom are known as *Cooper pair*. These pairs have zero total intrinsic angular momentum, all pair have same drift velocity and behaves like a boson.

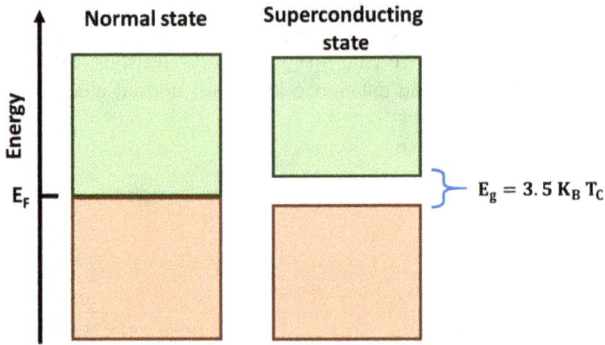

Figure 11.   *Band diagram for a superconductor in normal and superconducting state.*

In 1956, Rolfe Glover and Michael Tinkham studied the radiation absorption on a superconductor material exposed to infrared radiation (IR). It was observed that superconductor absorbed the IR only above a threshold value, below which absorption did not take place. As shown in Fig.11, a gap is created in the energy distribution due to the formation of a cooper pair that restricts their motion towards higher energy level. Due to the presence of this gap in the superconductors, scattering of electrons is not the only origin of current unlike in a normal conductor [1]. This energy or bandgap is the energy difference between free and paired electrons. At absolute zero, this gap is maximum and minimum to zero at critical temperature. At zero kelvin, it is directly proportional to the critical temperature of superconductor.

$$E_g = 3.5\, K_B\, T_C \tag{25}$$

Here, K_B is the Boltzmann constant and T_C is the critical temperature. Finally, superconductivity is defined with three phenomena, infinite conductivity, perfect diamagnetism, and the existence of energy bandgap.

5.7 Breakthroughs and outcomes of theoretical research

- The phenomenological theory of London brothers introduced the concept of penetration depth (λ) and successfully explained the perfect diamagnetism and perfect conductivity.

- Ginzburg and Landau theory introduced the concept of coherence length (ξ) to define a characteristic distance scale of superconductivity. The ratio of penetration depth and coherence length distinguish between type-I and type-II superconductors.

- BCS theory explains the electron-lattice-electron interaction, cooper pair formation and the existence of allowed energy states.

6. Applications

Superconductors have various applications depending on the specific properties as shown in Fig. 12. They are widely used for medical imaging in Nuclear Magnetic Resonance (NMR) [19,20] to scan tumours, brain activity and defective cells. Such a strong magnetic field is used to identify magnetic and non-magnetic materials. Further, it is possible to differentiate between a healthy and diseased cell tissues and organ in a human body. Superconductors are used in particle accelerators to generate a strong and stable magnetic field [21]. That is required to focus and maintain the path of particles or beam in a long synchrotron particle accelerator. Due to low energy loss, SCs are used to fabricate electric generators with high efficiency than conventional generators [22]. Electronic sensors and SQUIDs (superconducting quantum interference devices) are very sensitive devices that can detect and measure low magnetic fields. Basic mechanism of detector depends on the fact that even a small variation of voltage in superconducting coils can produce a large amount of constant current [23]. Superconductivity in type-I SCs can be reversed and destroyed easily hence, used for the fabrication of fast switching devices [24]. They are also used in memory devices due to their property of persistence current for a long period of time. Magnetic levitation and use of frictionless bearings in Mag Lev train is an example of application of SCs for transportation [25,26].

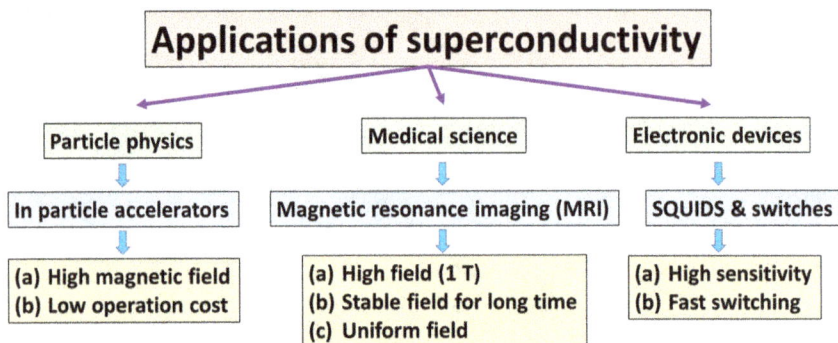

Figure 12. Applications of superconductivity in various fields.

7. Issues with type-I superconductors

Type-I superconductors have very low critical temperature (T_C) of a few Kelvin. Achieving and maintaining that temperature is not feasible. Mercury has a transition temperature of 4.2 K, that can be achieved using liquid helium but using it for practical applications is not economical. Another problem is the low critical magnetic field. In the presence of external field, type-I superconductivity vanishes easily. Due to limiting value of applied currents in wires they cannot be used to make an electromagnet.

To overcome all these limitations researchers worked on a new class of superconductors which have sufficiently large critical temperature and can withstand large magnetic field. In 1930 lead-bismuth was the first alloy to exhibit superconductivity, followed by alloys discovered later. Barium-copper-oxide ceramics got attention due to their high temperature superconductivity. Several materials were known to exhibit superconductivity at low temperature but 23 K was the known T_C highest until the discovery of high temp superconductors in 1986. Achieving superconductivity at room temperature is a dream for science community. But researchers worked to achieve superconductivity at liquid nitrogen temperature i.e. 77K if room temperature could not possible. Storing liquid nitrogen and maintaining this temperature is more economic and convenient than liquid helium.

In this chapter we have skipped thermodynamic properties, flux quantization and quantum tunnelling intentionally, that will be covered in another chapter. We have discussed *type-II and high-temperature superconductors* in detail in another chapter in this book.

References

[1] S. J. Blundell, Superconductivity: A Very Short Introduction. OUP Oxford, 2009. https://doi.org/10.1093/actrade/9780199540907.001.0001

[2] B.T. Matthias, T.H. Geballe, V.B. Compton, Superconductivity, Rev. Mod. Phys. 35 (1963) 1. https://doi.org/10.1103/RevModPhys.35.1

[3] Heike Kamerlingh Onnes - Nobel Lecture: Investigations into the Properties of Substances at Low Temperatures, which Have Led, amongst Other Things, to the Preparation of Liquid Helium. "Physics Nobel prize 1913."

[4] Rohlf, J. W. "Wiley: Modern Physics from alpha to Z0-James William Rohlf." (1994).

[5] E. Kaxiras, Atomic and Electronic Structure of Solids, At. Electron. Struct. Solids. (2003). https://doi.org/10.1017/CBO9780511755545

[6] C. Buchal, F. Pobell, R.M. Mueller, M. Kubota, J.R. Owers-Bradley, Superconductivity of Rhodium at Ultralow Temperatures, Phys. Rev. Lett. 50 (1983) 64. https://doi.org/10.1103/PhysRevLett.50.64

[7] R.D. Fowler, B.T. Matthias, L.B. Asprey, H.H. Hill, J.D.G. Lindsay, C.E. Olsen, R.W. White, Superconductivity of Protactinium, Phys. Rev. Lett. 15 (1965) 860. https://doi.org/10.1103/PhysRevLett.15.860

[8] D.R. Lide, CRC handbook of chemistry and physics, CRC press, 2004.

[9] R.L. Falge, Superconductivity of hexagonal beryllium, Phys. Lett. A. 24 (1967) 579-580. https://doi.org/10.1016/0375-9601(67)90624-X

[10] J. Eisenstein, Superconducting Elements, Rev. Mod. Phys. 26 (1954) 277. https://doi.org/10.1103/RevModPhys.26.277

[11] G. Seidel, P.H. Keesom, Specific Heat of Gallium and Zinc in the Normal and Superconducting States, Phys. Rev. 112 (1958) 1083. https://doi.org/10.1103/PhysRev.112.1083

[12] J.F. Cochran, D.E. Mapother, Superconducting Transition in Aluminum, Phys. Rev. 111 (1958) 132. https://doi.org/10.1103/PhysRev.111.132

[13] K.I. Wysokinski, Remarks on the first hundred years of superconductivity, Acta Phys. Pol. A. 121 (2011) 721-725. https://doi.org/10.12693/APhysPolA.121.721

[14] E. Maxwell, Isotope Effect in the Superconductivity of Mercury, Phys. Rev. 78 (1950) 477. https://doi.org/10.1103/PhysRev.78.477

[15] C.A. Reynolds, B. Serin, W.H. Wright, L.B. Nesbitt, Superconductivity of Isotopes of Mercury, Phys. Rev. 78 (1950) 487. https://doi.org/10.1103/PhysRev.78.487

[16] The electromagnetic equations of the supraconductor, Proc. R. Soc. London. Ser. A - Math. Phys. Sci. 149 (1935) 71-88. https://doi.org/10.1098/rspa.1935.0048

[17] M. Cyrot, Ginzburg-Landau theory for superconductors, Reports Prog. Phys. 36 (1973) 103. https://doi.org/10.1088/0034-4885/36/2/001

[18] J. Bardeen, L.N. Cooper, J.R. Schrieffer, Theory of Superconductivity, Phys. Rev. 108 (1957) 1175. https://doi.org/10.1103/PhysRev.108.1175

[19] T. Nakamura, Y. Itoh, M. Yoshikawa, T. Oka, J. Uzawa, Development of a superconducting magnet for nuclear magnetic resonance using bulk high-temperature superconducting materials, Concepts Magn. Reson. Part B Magn. Reson. Eng. 31B (2007) 65-70. https://doi.org/10.1002/cmr.b.20083

[20] K. Ogawa, T. Nakamura, Y. Terada, K. Kose, T. Haishi, Development of a magnetic resonance microscope using a high Tc bulk superconducting magnet, Appl. Phys. Lett. 98 (2011) 234101. https://doi.org/10.1063/1.3598440

[21] K. Takahashi, H. Fujishiro, M.D. Ainslie, A new concept of a hybrid trapped field magnet lens, Supercond. Sci. Technol. 31 (2018) 044005. https://doi.org/10.1088/1361-6668/aaae94

[22] H. Fujishiro, M. Ikebe, H. Teshima, H. Hirano, Low-thermal-conductive DyBaCuO bulk superconductor for current lead application, IEEE Trans. Appl. Supercond. 16 (2006) 1007-1010. https://doi.org/10.1109/TASC.2006.871305

[23] H. Deng, Y. Wu, Y. Zheng, N. Akhtar, J. Fan, X. Zhu, J. Li, Y. Jin, D. Zheng, Working Point Adjustable DC-SQUID for the Readout of Gap Tunable Flux Qubit, IEEE Trans. Appl. Supercond. 25 (2015). https://doi.org/10.1109/TASC.2015.2399272

[24] T. Polakovic, W.R. Armstrong, V. Yefremenko, J.E. Pearson, K. Hafidi, G. Karapetrov, Z.E. Meziani, V. Novosad, Superconducting nanowires as high-rate photon detectors in strong magnetic fields, Nucl. Instruments Methods Phys. Res. Sect. A Accel. Spectrometers, Detect. Assoc. Equip. 959 (2020) 163543. https://doi.org/10.1016/j.nima.2020.163543

[25] L. Schultz, O. De Haas, P. Verges, C. Beyer, S. Röhlig, H. Olsen, L. Kühn, D. Berger, U. Noteboom, U. Funk, Superconductively levitated transport system-the SupraTrans project, IEEE Trans. Appl. Supercond. 15 (2005) 2301-2305. https://doi.org/10.1109/TASC.2005.849636

[26] G.G. Sotelo, D.H.N. Dias, O.J. Machado, E.D. David, R. de A. Jr, R.M. Stephan, G.C. Costa, Experiments in a real scale maglev vehicle prototype, J. Phys. Conf. Ser. 234 (2010) 032054. https://doi.org/10.1088/1742-6596/234/3/032054

Superconductors - Materials and Applications
Materials Research Foundations 132 (2022) 131-145

Materials Research Forum LLC
https://doi.org/10.21741/9781644902110-7

Chapter 7

Bulk Superconductors: Materials and Applications

Prashant Hitaishi[1], Rohit Verma[2]*, Parul Khurana[3] and Sheenam Thatai[2]

[1]Department of Physics, School of Natural Sciences, Shiv Nadar University, G.B. Nagar, 201314, India

[2]Amity Institute of Applied Sciences, Amity University, Noida, 201313, India

[3]G.N.Khalsa College, Mumbai University, Mumbai, 400019, India

* rverma85@amity.edu

Abstract

Bulk or high temperature superconductors are materials that are not perfect diamagnetic, but attain zero resistivity or infinite conductivity below a well-defined temperature known as critical temperature (T_C). In the presence of external magnetic field, such material exhibit two critical field values H_{C1} and H_{C2}. These materials may exist in three different states: superconducting, vortex or mixed and normal state depending upon the intensity of the applied field. The principal application of such SCs is in compact strong and stable field permanent magnets, and the magnetic flux density of these magnets is much higher than conventional permanent magnets. This chapter has discussed various types of high-temperature superconductors, their physical properties, applications, and recent research in the field.

Keywords

Superconductors, Type-II, Bulk or High Temperature Superconductors, Vortex State, Oxide, Cuprates, High Field Magnets, Applications of Superconductor

Contents

1.   Introduction

Since 1911, after the discovery of superconductivity it has been an active field of research. Initially, superconductivity was observed and studied only in few pure-metals known as type-I superconductors (SCs). The main requirement of very low temperature made it difficult and expensive to use them for potential applications. As a result, people started to look for economic and feasible alternatives to achieve high-temperature superconductivity. That was the beginning of a new era in which a lot of research work has been done on various materials like metals, alloys, ceramics, interstitial and intermetallic compounds. Niobium (Nb), one of the exceptional elements that exhibit type-II superconducting nature, showed a high critical temperature of 9.25 K in 1930. In 1930, W. J. de Haas, J. Voogd and W. H. Keesom at Leiden laboratory discovered superconductivity in lead-bismuth alloy [1].

In 1977, an intermetallic compound consisting of germanium and niobium (Nb3Ge) achieved the maximum critical temperature of 23 K. Interestingly, elements used in the alloy did not show superconductivity. Still, alloy showed an enhanced critical magnetic field. It was a great success in this field with new hope and direction for research.

2.   New era of high temperature superconductor

Several materials were known to exhibit superconductivity at low temperature and 23.2 K in Nb_3Ge was the known T_C highest until the discovery of high temp SCs in 1986 [1]. Till 1960s various alloys of niobium (Nb) were discovered that behaves as a superconductor at 10-23 K. And a general perception was established on theoretical ground that 30 K could be the highest achievable critical temperature [2]. Bednorz and Muller discovered La-Ba-Cu-O, a ceramic superconductor that showed T_C of 30 K in 1986 [3] and finally, this discovery of such a promising oxide ceramic superconductor was awarded the Noble Prize in 1987. It was surprising to note that ceramic, being an insulator showed superconductivity with a high value of T_C. In 1987, a ceramic superconductor $YBa_2Cu_3O_7$ attain T_C of 90K and in 1988 a cuprate superconductor named thallium cuprates ($Tl_2Ba_2Cu_3O_{10}$) showed T_C of 125K. Materials with T_C in the range of 120 K received more attention because it is convenient to achieve and maintain this temperature using liquid nitrogen (77 K).

3.   Type-II superconductors

Vanadium, technetium, and niobium are three exception elements that are type-II superconductors. Initially, they were listed in thirty pure metal type-I superconductors but later removed from the list. Generally, type-II superconductors are alloys, complex oxides, and ceramics. They are known for their superconductive behavior at relatively high temperatures than type-I SCs. Induced current in high temperature superconductors (HTS) or type-II SCs do not vanish even in the absence of active power source due to low or almost zero resistivity of material [4].

4.   Characteristics of type-II superconductors

4.1   Critical temperature (T_C)

The phase transition from normal to superconducting state occurred at a particular temperature known as *critical temperature* (T_C). Type-II SCs do not show an abrupt or sudden transition in the properties at T_C and follow a smooth transition. They have relatively higher transition temperature than type-I SCs. List of various type-II SCs is tabulated in table 1. They exists in three different states like normal, mixed and superconducting depending upon the temperature and externally applied magnetic field. These states define a range or boundary for a particular material known as magnetic phase diagram for type-II superconductors is shown in Fig. 1.

Table 1:   *List of type-II and bulk superconductors and their critical transition temperature. Tabulated as year-wise development in the materials.*

Material	Year	T_C	Reference
Niobium	1930	9.2	[9]
Lead-thallium (PbTl$_2$)	1935	3.75	[7]
Niobium-nitride	1941	16	[33]
Vanadium-silicon	1954	17.1	[34]
Niobium-titanium (NbTi) *First commercial superconducting wire*	1962	10	[35]
Mo$_{6-x}$A$_x$S$_6$ (here A is Mg, Ag, Pb, Sn, Cu, Zn or Cd) *First ternary compound*	1972	2.5K (for Cd) to 13K (for Pb)	[5]
Li$_x$Ti$_{1.1}$S$_2$	1972	10 to 13	[6]
LiTiO	1973	7 to 13.7	[15]
BaPb$_{1-x}$Bi$_x$O$_3$	1975	13 K	[16]
Nb$_3$Ge	1977	23K	
(TMTSF)$_2$PF$_6$	1980	0.9 K	[36]
LaBaCuO	1986	30K	[3]
LaSrCuO	1981	40 K	[37]
BiSrCuO (BSCO)	1987	22 K	[21]
YBa$_2$Cu$_3$O$_7$	1987	80 to 93 K	[37]
(Bi,Pb)SrCaCuO (BSCCO)	1997	105 K	[22,38]
Thallium cuprates (TlBaCaCuO)	1988	118 to 125K	[23]
Caesium-doped C$_{60}$ (Cs$_x$C60)	1991	30 K	[26]
Potasssium-doped C$_{60}$	1991	18 K	[25]

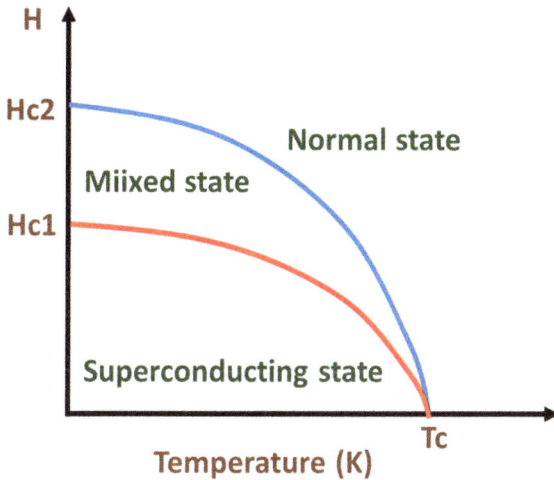

Figure 1:   Magnetic phase diagram for type-II superconductors

4.2   Critical magnetic field (H_C)

Type-II SCs do not lose their superconductivity abruptly in the presence of an external magnetic field and exhibits two critical fields. They exhibit higher critical magnetic field and current density than type-I SCs. They gradually lose superconductivity in the presence of externally applied magnetic field but not easily and are mechanically harder than type-I SCs hence known as hard superconductors. Unlike type-I SCs there are two critical magnetic fields (H_C). At lower critical magnetic field (H_{C1}), type-II SCs starts to lose their superconductivity and completely lose it at the upper critical field (H_{C2}). Above H_{C2} a type-II SC behaves like a normal conductor. The intermediate state that exists between H_{C1} and H_{C2} is a mixed state of the normal and superconducting state hence known as intermediate, mixed or vortex state. Magnetization (M) of type-II superconductors in response to the externally applied magnetic field (H). And modification in the vortices with the increase in an applied magnetic field is shown in Fig. 2. As shown in Fig. 3, these vortices consist of core and circulating supercurrents in the surrounding. Generally, the size of normal cores is 300 nm, but in the presence of an externally applied strong magnetic field, vortices come close and start overlapping each other. Type-II SCs can transport current with zero resistance even in the mixed state and until the vortices are stationary. Once vortices start moving and come close, cores overlap and material loses its superconductivity [1]. Vortices feel a repulsive force in the absence of flow of current in the sample while an attractive force is experienced in the presence of current in the sample that causes the movement in vortices. They have a wider range of applications due to their high critical magnetic field.

If a superconductor is placed in an externally applied magnetic field intensity H then, B is magnetic flux density inside it and M is magnetization. We can write,

$$\vec{B} = \mu_0(\vec{H} + \vec{M}) \tag{1}$$

As shown in Fig. 2 in a type-II SC magnetic field can penetrate the material within a range Hc1 < H < Hc2 hence, the external field is not completely canceled by the magnetization. At Hc2, normal cores of vortices overlap each other as a result, magnetization is zero, superconductivity vanishes and a normal state is recovered [1].

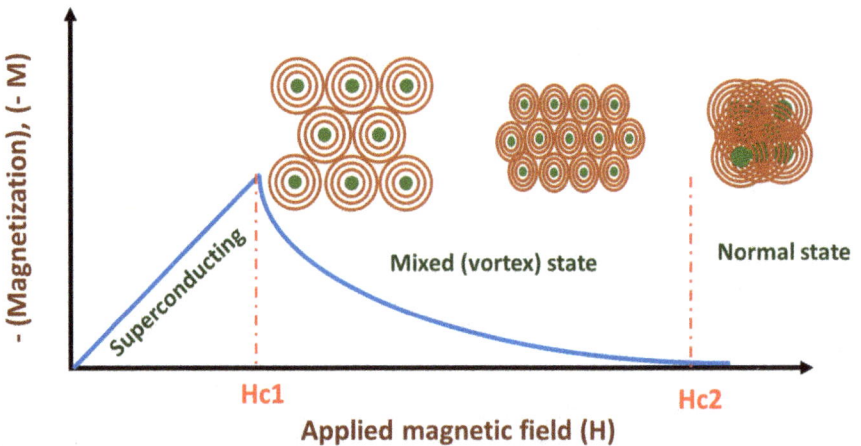

Figure 2:   *Magnetization (M) of type-II superconductors in response to the externally applied magnetic field (H). And modification in the vortices with the increase in applied magnetic field.*

4.3   Meissner effect or perfect diamagnetism

Type-II SCs are not perfect diamagnetic material hence do not follow the Meissner effect completely. Bulk superconductors are used as *trapped field magnets* because they exhibit a strong and stable magnetic field of several teslas when cooled below its critical temperature. Penetration of magnetic flux into the material starts at the lower critical field (H_{C1}) and increases with the increase in an externally applied field. Finally, beyond upper critical field (H_{C2}), magnetization fully penetrates into the material and loses superconductivity completely [2]. For type-II SCs coherence length (ξ) is less than penetration depth (λ) and both are related by relation $\xi < \sqrt{2}\lambda$. Properties of type-II SCs get changed even in the presence of slight impurity while there is hardly any effect on type-I SCs

5. Different types of bulk superconductors

5.1 Alloys

The first ternary high-temperature superconductor reported in 1972 was $Mo_{6-x}A_xS_6$ (here A is Mg, Ag, Pb, Sn, Cu, Zn or Cd). The transition temperature varies from 2.5K (for Cd) to 13K (for Pb) [5]. Lithium titanium sulphide ($Li_xTi_{1.1}S_2$) was the first compound of noncubic class of superconductors reported in 1972 and showed superconductivity from 10 to 13K [6]. The superconducting nature of alloy of lead-thallium ($PbTl_2$) has been reported in [7]. And its critical temperature was determined by electrical and magnetic property measurement hence, it comes out to be 3.75K.

5.2 Niobium alloys

Niobium is a metal that gained much attention and application in advanced technologies due to its high melting point (2468 °C), superconductivity, dielectric oxides formation and corrosion resistance. But all these properties strongly depend on the purity of niobium [8]. B.T. Matthias and his collaborators measured the critical temperature (18 K) of niobium-tin (Nb_3Sn) in 1953 [1]. It is widely used to construct superconducting radiofrequency cavities for applications in particle accelerators. It has critical temperature $T_C = 9.2$ K and field $B_C = 196$ mT while its alloy named Nb_3Sn showed enhanced properties with $T_C = 18.2$ K and field $B_C = 535$ mT [9]. Electrical and superconducting properties of a thin film of niobium nitride (NbN) have been studied to understand the close relation between film structure and critical temperature. Variation in thin film sputtering conditions and substrate showed an unexpected change in T_C. An increase in film thickness from 1000 to 8000 Å increased the T_C by 1.5K [10]. Superconducting wires of niobium alloys like niobium-tin (Nb_3Sn), niobium-titanium (Nb-Ti) and niobium-germanium (Nb_3Ge) are used in superconducting magnets. These magnets can withstand upto a very high magnetic field of few teslas. The magnetic field produced by these magnets depends on the material, amount of superconducting wire used in the winding of coils, and magnet size. These super magnets have various applications like particle accelerators, nuclear magnetic resonance (NMR) and magnetic resonance imaging (MRI) [11–13]. Niobium nitride (NbN) thin-film exhibit excellent superconducting properties with $T_C = 16$ K [14].

5.3 Oxides, cuprates and ceramics

$Li_{1+x}Ti_{2-x}O_4$ was the first high temperature oxide superconductor reported in 1973, showed the transition temperature range from 7 to 13.7 K [15]. Various phases of type $BaPb_{1-x}Bi_xO_3$ were prepared first time in 1975, and this compound showed the highest $T_C = 13$ K [16]. (RE)BCO (here RE is rare earth) is a material with an increased capability to trap a magnetic field in it. It is widely used to produce high field permanent magnets that can generate strong field of 2 T (in a bulk sample of YBCO of 20 mm diameter) [17] and 3 T

(65 mm diameter sample of GdBaCuO) [18] at 77 K. Bulk YBaCuO sample of 2.65 cm diameter with some modifications in magnet design could trap field of 17.24 T at 29 K [19]. This sample showed enhanced mechanical strength and thermal stability without fracture in it. And the trapped field of 17.6 T was the maximum till 2014, achieved in a silver-doped GdBCo sample of 25 mm diameter at 26 K [20]. Another superconducting oxide, a novel family BiSrCuO showed transition at temperatures 22 K and 7 K for ultrapure oxide and commercial compound respectively. X-ray diffraction and electron microscopy have been used to study the structure of this compound [21]. $(LaSr)_2CuO_4$ and $YBa_2Cu_3O_7$ have high $T_C = 105$ K and are oxides that are highly stable in moisture, water. There is no degradation in the sample after multiple cycles between 4 K and room temperature hence due to their high stability, they have potential industrial applications [22].

In 1988, a cuprate superconductor named thallium cuprate ($Tl_2Ba_2Cu_3O_{10}$) showed T_C of 125 K [23] and produced a field of 5 T at 4.3 K [24]. Materials with T_C in the range of liquid nitrogen got more attention because it is convenient to achieve and maintain this temperature using liquid nitrogen (77 K). The fundamental and geometrical configuration of high-temperature superconductor is composed of copper-oxides layer with other atoms that acts as reservoirs and spacers. In the material LaBaCuO, lanthanum-barium layers are used as spacer layers for copper-oxide layers. The critical temperature of a material and its sensitivity strongly depend on the spacer atoms. La_2CuO_4 is not a superconductor but just with the replacement of few atoms (~10%) of lanthanum by barium made it a superconductor. Strontium and calcium are other dopants and alternatives to barium that can be used for the same purpose [1]. CuO layered plane is a common feature present in all cuprates.

5.4   Fullerenes

Masami Tanigaki demonstrated first-time fullerene superconductor in 1991. Its reported critical temperature was 33 K, the highest for a molecular superconductor. Then another groups discovered the superconductivity in potassium-doped C_{60} ($T_C = 18K$) [25] and caesium-doped C_{60} ($T_C = 30K$) [26]. The transition temperature of A_3C_{60} (A = alkali metals) depends on the inter C_{60} separation, which depends on the size of cation A^+. Modeling of structure Cs_3C_{60}, bulk superconductor helped to attain the $T_C = 38$ K in a molecular system [25].

6.   Applications

6.1   Superconductor magnets and ordinary electromagnets

Superconductor has zero or negligible resistance in the superconducting state hence, it carries the current for years without any loss. The flow of this persistent current only requires a low temperature to keep the material in a superconducting state. It implies the running cost of such a system is only refrigeration cost, which comes out to be less than the operation cost of an ordinary electromagnet (OEM). If copper wire is used to build an

OEM, it requires a continuous power source because electricity in the wire is lost due to its resistance. Hence, the overall cost required to run a refrigerating system is less than the amount wasted in the form of electricity in OEM that made a superconductor magnet cost-effective [2].

6.2   High field magnets

The magnetic field produced by super magnets is very high as compared to conventional permanent or electromagnets. As a result, they are widely used for various engineering applications like high field magnets, compact, high power density, and efficient electrical machines. These applications made it possible to understand the dependency of physical and chemical properties of the material on magnetism. Such materials have revolutionized our life due to numerous applications in medical science. Nuclear magnetic resonance (NMR), magnetic resonance imaging (MRI), and magnetic drug delivery systems (MDDS) are some potential examples [27]. A comparison between the magnetic field produced by a permanent ferromagnetic, electromagnet and bulk superconductor is shown in Fig.3. The magnetic field produced by a permanent magnet is of low intensity but uniform. An electromagnet wounded by a normal conductor or superconductor wire requires an externally supplied loop current in the coils. While in a superconductor magnet, no power supply or continuous current source is required because loop currents are self-induced in the superconducting state. A comparison of the origin of magnetism in three different types of magnets and variation in the magnetic field is shown in Fig. 3.

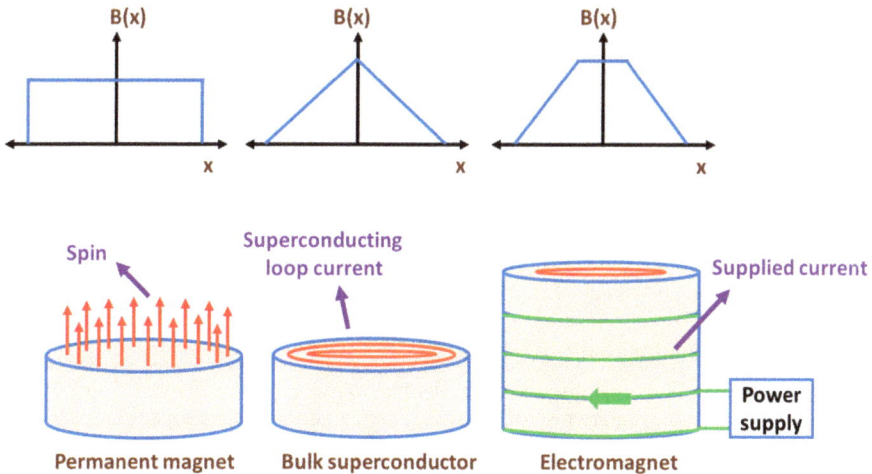

Figure 3:   A comparison of origin of magnetism in three different type of magnets and variation in the magnetic flux as a function of distance measured from the centre of magnet.

6.3   Magnetic levitation

The magnetic levitation (MagLev) in superconductors is different from permanent magnets as it is caused due to the supercurrents inside the superconductor. The Meissner effect is so strong that a magnet can be levitated over a superconductor. A small volume of such material is capable to produce a magnetic field of several Tesla and also the levitation caused is self-stabilised. MagLev transport system allows a vehicle to levitate over the track with minimal friction [27].

6.4   Medical applications

The ability of superconducting magnets to generate a strong, stable, and uniform magnetic field at a low operational cost made it possible to use them for medical applications. In magnetic resonance imaging (MRI) and Nuclear Magnetic Resonance (NMR) to scan tumors, brain activity, and defective cells a strong magnetic field of few teslas is required for a long duration. MRI works on the principle that the magnetic moment of a hydrogen nucleus precesses at a particular frequency in the presence of an externally applied electromagnetic field. Portable MRI systems are possible because of compact and high field magnets.

6.5   Detectors

Niobium nitride (NbN) superconducting wires are used in photons [28] and particle detectors. Such devices can detect photons of wavelength 400 nm up to a strong magnetic field of 5 T [29]. Superconducting quantum interference devices (SQUIDs) are sensitive magnetic flux detectors that measure a voltage very precisely. Superconducting magnets are used in particle accelerators and spectrometers because of their strong magnetic field and low operation cost.

6.6   Josephson junctions

These are used in electronic devices for fast switching applications due to their low power consumption and very quick switch response from normal to superconducting state or vice versa. The switching time for an energy gap of 1 meV is roughly 4 ps, with around few microwatts power dissipation [1]. More applications of bulk SCs are shown in Fig. 4.

Conclusion and future outlook

In the above sections, we discussed various type of superconductors and their potential applications. But the requirement of very cold temperature and high pressure is still a major concern that limits various immediate practical applications. Researchers are working worldwide to develop new materials that can be used even at room temperature and atmospheric pressure. Different groups studied superconductivity in hydrogen-rich materials and metallic hydrogen compounds that shows superconductivity close to room temperature but require high pressure. Lanthanum superhydride show superconductivity at 260 K [30] and megabar pressure, sulfur hydride at 203 K [31-32]. The future perspective of superconductivity includes development in materials to attain room temperature superconductivity. Various applications like strong, stable, lightweight and more efficient magnets to develop portable devices that will revolutionize our lives.

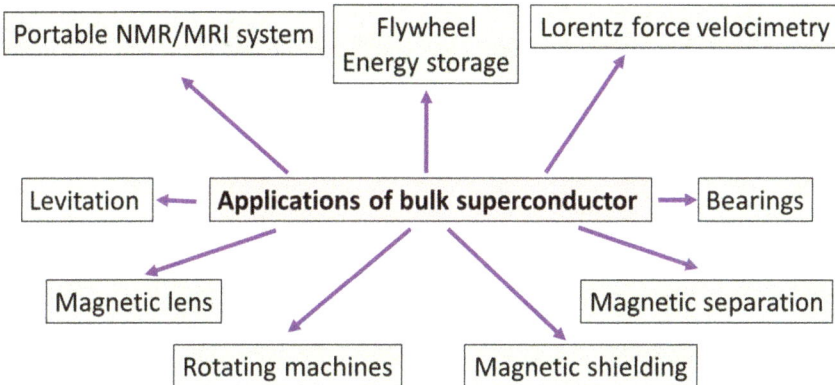

Figure 4:   *Various applications of bulk superconductor*

Reference

[1] Rohlf, J. W. "Wiley: Modern Physics from alpha to Z0-James William Rohlf." (1994).

[2] T.P. Sheahen, Introduction to High-Temperature Superconductivity, Introd. to High-Temperature Supercond. (2002). https://doi.org/10.1007/B115100. https://doi.org/10.1007/b115100

[3] J.G. Bednorz, K.A. Müller, Possible highT c superconductivity in the Ba−La−Cu−O system, Zeitschrift Für Phys. B Condens. Matter 1986 642. 64 (1986) 189-193. https://doi.org/10.1007/BF01303701. https://doi.org/10.1007/BF01303701

[4] E. Kurbatova, P. Kurbatov, M. Sysoev, Bulk High-Temperature Superconductors: Simulation of Electromagnetic Properties, Prop. Nov. Supercond. (2020). https://doi.org/10.5772/INTECHOPEN.92452. https://doi.org/10.5772/intechopen.92452

[5] B.T. Matthias, M. Marezio, E. Corenzwit, A.S. Cooper, H.E. Barz, High-Temperature Superconductors, the First Ternary System, Science (80-.). 175 (1972) 1465-1466. https://doi.org/10.1126/SCIENCE.175.4029.1465. https://doi.org/10.1126/science.175.4029.1465

Materials Research Forum LLC
https://doi.org/10.21741/9781644902110-7

[6] H.E. Barz, A.S. Cooper, E. Corenzwit, M. Marezio, B.T. Matthias, P.H. Schmidt, Superconductivity of Double Chalcogenides: Lix Ti1.1S2, Science (80-.). 175 (1972) 884-885. https://doi.org/10.1126/SCIENCE.175.4024.884. https://doi.org/10.1126/science.175.4024.884

[7] J.N. RJABININ, L.W. SHUBNIKOW, Magnetic Properties and Critical Currents of Supra-conducting Alloys, Nat. 1935 1353415. 135 (1935) 581-582. https://doi.org/10.1038/135581a0. https://doi.org/10.1038/135581a0

[8] H.R.S. Moura, L. de Moura, Melting And Purification Of Niobium, AIP Conf. Proc. 927 (2007) 165. https://doi.org/10.1063/1.2770689. https://doi.org/10.1063/1.2770689

[9] M. Peiniger, H. Piel, A superconducting nb3sn coated multicell accelerating cavity, IEEE Trans. Nucl. Sci. 32 (1985) 3610-3612. https://doi.org/10.1109/TNS.1985.4334443. https://doi.org/10.1109/TNS.1985.4334443

[10] Y.M. Shy, L.E. Toth, R. Somasundaram, Superconducting properties, electrical resistivities, and structure of NbN thin films, J. Appl. Phys. 44 (2003) 5539. https://doi.org/10.1063/1.1662193. https://doi.org/10.1063/1.1662193

[11] G. Grunblatt, P. Mocaer, C. Verwaerde, C. Kohler, A success story: LHC cable production at ALSTOM-MSA, Fusion Eng. Des. 75-79 (2005) 1-5. https://doi.org/10.1016/J.FUSENGDES.2005.06.216. https://doi.org/10.1016/j.fusengdes.2005.06.216

[12] B.A. Glowacki, X.Y. Yan, D. Fray, G. Chen, M. Majoros, Y. Shi, Niobium based intermetallics as a source of high-current/high magnetic field superconductors, Phys. C Supercond. 372-376 (2002) 1315-1320. https://doi.org/10.1016/S0921-4534(02)01018-3. https://doi.org/10.1016/S0921-4534(02)01018-3

[13] J.L.H. Lindenhovius, E.M. Hornsveld, A. Den Ouden, W.A.J. Wessel, H.H.J. Ten Kate, PowderinTube (PIT) Nb3Sn conductors for highfield magnets, IEEE Trans. Appl. Supercond. 10 (2000) 975978. https://doi.org/10.1109/77.828394. https://doi.org/10.1109/77.828394

[14] Z. Wang, A. Kawakami, Y. Uzawa, B. Komiyama, Superconducting properties and crystal structures of single-crystal niobium nitride thin films deposited at ambient substrate temperature, J. Appl. Phys. 79 (1998) 7837. https://doi.org/10.1063/1.362392. https://doi.org/10.1063/1.362392

[15] D.C. Johnston, H. Prakash, W.H. Zachariasen, R. Viswanathan, High temperature superconductivity in the LiTiO ternary system, Mater. Res. Bull. 8 (1973) 777-784. https://doi.org/10.1016/0025-5408(73)90183-9. https://doi.org/10.1016/0025-5408(73)90183-9

[16] A.W. Sleight, J.L. Gillson, P.E. Bierstedt, High-temperature superconductivity in the BaPb1-xBixO3 systems, Solid State Commun. 17 (1975) 27-28.

Superconductors - Materials and Applications   Materials Research Forum LLC
Materials Research Foundations 132 (2022) 131-145   https://doi.org/10.21741/9781644902110-7

https://doi.org/10.1016/0038-1098(75)90327-0. https://doi.org/10.1016/0038-1098(75)90327-0

[17] R.-P. Sawh, R. Weinstein, K. Carpenter, D. Parks, K. Davey, Production run of 2 cm diameter YBCO trapped field magnets with surface field of 2 T at 77 K, Supercond. Sci. Technol. 26 (2013) 105014. https://doi.org/10.1088/0953-2048/26/10/105014. https://doi.org/10.1088/0953-2048/26/10/105014

[18] S. Nariki, N. Sakai, M. Murakami, Melt-processed Gd-Ba-Cu-O superconductor with trapped field of 3 T at 77 K, Supercond. Sci. Technol. 18 (2004) S126. https://doi.org/10.1088/0953-2048/18/2/026. https://doi.org/10.1088/0953-2048/18/2/026

[19] M. Tomita, M. Murakami, High-temperature superconductor bulk magnets that can trap magnetic fields of over 17 tesla at 29 K, Nat. 2003 4216922. 421 (2003) 517-520. https://doi.org/10.1038/nature01350. https://doi.org/10.1038/nature01350

[20] J.H. Durrell, A.R. Dennis, J. Jaroszynski, M.D. Ainslie, K.G.B. Palmer, Y.-H. Shi, A.M. Campbell, J. Hull, M. Strasik, E.E. Hellstrom, D.A. Cardwell, A trapped field of 17.6 T in melt-processed, bulk Gd-Ba-Cu-O reinforced with shrink-fit steel, Supercond. Sci. Technol. 27 (2014) 082001. https://doi.org/10.1088/0953-2048/27/8/082001. https://doi.org/10.1088/0953-2048/27/8/082001

[21] C. Michel, M. Hervieu, M.M. Borel, A. Grandin, F. Deslandes, J. Provost, B. Raveau, Superconductivity in the Bi - Sr - Cu - O system, Zeitschrift Für Phys. B Condens. Matter. 68 (1987) 421-423. https://doi.org/10.1007/BF01471071. https://doi.org/10.1007/BF01471071

[22] W. Miller, K. Borówko, M. Gazda, S. Stizza, R. Natali, Proceedings of the XI National School "Collective Phenomena and Their Competition, Kazimierz Dolny. 109 (2006). https://doi.org/10.12693/APhysPolA.109.627

[23] S.S.P. Parkin, V.Y. Lee, E.M. Engler, A.I. Nazzal, T.C. Huang, G. Gorman, R. Savoy, R. Beyers, Bulk Superconductivity at 125 K in <span class, Phys. Rev. Lett. 60 (1988) 2539. https://doi.org/10.1103/PhysRevLett.60.2539. https://doi.org/10.1103/PhysRevLett.60.2539

[24] S. ying Ding, Z. Yu, L. Qiu, Magnetic characters of polycrystalline TlBaCaCuO superconductor, Phys. B Condens. Matter. 165-166 (1990) 1407-1408. https://doi.org/10.1016/S0921-4526(09)80289-5. https://doi.org/10.1016/S0921-4526(09)80289-5

[25] A.F. Hebard, M.J. Rosseinsky, R.C. Haddon, D.W. Murphy, S.H. Glarum, T.T.M. Palstra, A.P. Ramirez, A.R. Kortan, Superconductivity at 18 K in potassium-doped C60, Nat. 1991 3506319. 350 (1991) 600-601. https://doi.org/10.1038/350600a0. https://doi.org/10.1038/350600a0

Materials Research Forum LLC
https://doi.org/10.21741/9781644902110-7

[26] S.P. Kelty, C.-C. Chen, C.M. Lieber, Superconductivity at 30 K in caesium-doped C60, Nat. 1991 3526332. 352 (1991) 223-225. https://doi.org/10.1038/352223a0. https://doi.org/10.1038/352223a0

[27] M. Ainslie, H. Fujishiro, Fundamentals of bulk superconducting materials, Numer. Model. Bulk Supercond. Magn. (2019). https://doi.org/10.1088/978-0-7503-1332-2CH1. https://doi.org/10.1088/978-0-7503-1332-2ch1

[28] M. Hajenius, J.J.A. Baselmans, J.R. Gao, T.M. Klapwijk, P.A.J. de Korte, B. Voronov, G. Gol'tsman, Low noise NbN superconducting hot electron bolometer mixers at 1.9 and 2.5 THz, Supercond. Sci. Technol. 17 (2004) S224. https://doi.org/10.1088/0953-2048/17/5/026. https://doi.org/10.1088/0953-2048/17/5/026

[29] T. Polakovic, W.R. Armstrong, V. Yefremenko, J.E. Pearson, K. Hafidi, G. Karapetrov, Z.E. Meziani, V. Novosad, Superconducting nanowires as high-rate photon detectors in strong magnetic fields, Nucl. Instruments Methods Phys. Res. Sect. A Accel. Spectrometers, Detect. Assoc. Equip. 959 (2020) 163543. https://doi.org/10.1016/J.NIMA.2020.163543. https://doi.org/10.1016/j.nima.2020.163543

[30] M. Somayazulu, M. Ahart, A.K. Mishra, Z.M. Geballe, M. Baldini, Y. Meng, V. V. Struzhkin, R.J. Hemley, Evidence for Superconductivity above 260 K in Lanthanum Superhydride at Megabar Pressures, Phys. Rev. Lett. 122 (2019) 027001. https://doi.org/10.1103/PHYSREVLETT.122.027001/FIGURES/4/MEDIUM. https://doi.org/10.1103/PhysRevLett.122.027001

[31] A.P. Drozdov, M.I. Eremets, I.A. Troyan, V. Ksenofontov, S.I. Shylin, Conventional superconductivity at 203 kelvin at high pressures in the sulfur hydride system, Nat. 2015 5257567. 525 (2015) 73-76. https://doi.org/10.1038/nature14964. https://doi.org/10.1038/nature14964

[32] E. Snider, N. Dasenbrock-Gammon, R. McBride, M. Debessai, H. Vindana, K. Vencatasamy, K. V. Lawler, A. Salamat, R.P. Dias, Room-temperature superconductivity in a carbonaceous sulfur hydride, Nature. 586 (2020) 373-377. https://doi.org/10.1038/S41586-020-2801-Z. https://doi.org/10.1038/s41586-020-2801-z

[33] M. Khorrami, Mona, Superconductors (History & Advanced Research), APS. 2012 (2012) C1.300. https://ui.adsabs.harvard.edu/abs/2012APS..MAR.C1300K/abstract (accessed September 1, 2021).

[34] G.F. Hardy, J.K. Hulm, The Superconductivity of Some Transition Metal Compounds, Phys. Rev. 93 (1954) 1004. https://doi.org/10.1103/PhysRev.93.1004. https://doi.org/10.1103/PhysRev.93.1004

Superconductors - Materials and Applications   Materials Research Forum LLC
Materials Research Foundations 132 (2022) 131-145   https://doi.org/10.21741/9781644902110-7

[35] Z. Charifoulline, Residual Resistivity Ratio (RRR) measurements of LHC superconducting NbTi cable strands, IEEE Trans. Appl. Supercond. 16 (2006) 1188-1191. https://doi.org/10.1109/TASC.2006.873322. https://doi.org/10.1109/TASC.2006.873322

[36] D. Jérome, A. Mazaud, M. Ribault, K. Bechgaard, Superconductivity in a synthetic organic conductor (TMTSF)2PF 6, J. Phys. Lettres. 41 (1980) 95-98. https://doi.org/10.1051/JPHYSLET:0198000410409500. https://doi.org/10.1051/jphyslet:0198000410409500

[37] L. Er-Rakho, C. Michel, J. Provost, B. Raveau, A series of oxygen-defect perovskites containing CuII and CuIII: The oxides La3−xLnxBa3 [CuII5−2y CuIII1+2y] O14+y, J. Solid State Chem. 37 (1981) 151-156. https://doi.org/10.1016/0022-4596(81)90080-3. https://doi.org/10.1016/0022-4596(81)90080-3

[38] Y. Abe, Superconducting Glass-Ceramics In Bisrcacuo, Supercond. Glas. Bisrcacuo. (1997). https://doi.org/10.1142/3537. https://doi.org/10.1142/3537

Superconductors - Materials and Applications   Materials Research Forum LLC
Materials Research Foundations 132 (2022) 146-165   https://doi.org/10.21741/9781644902110-8

Chapter 8

Soft Superconductors: Materials and Applications

M. Bugdayci[1,3]*, S. Yesiltepe[2]

[1]Yalova University, Engineering Faculty Chemical Engineering Dept., Yalova/Turkey

[2]Korkutata University, Engineering Faculty, Mechanical Engineering Dept., Osmaniye/Turkey

[3]Istanbul Medipol University, Vocational School, Construction Technology Dep., 34810,Istanbul, Turkey

* mehmet.bugdayci@yalova.edu.tr

Abstract

Superconductors emerge as materials that offer great benefits with the superior properties they offer. In this study, the general properties of superconducting materials, their history and properties of soft superconductors are emphasized. Superconductivity mechanisms are investigated through crystal structures and BCS theory. Superconductivity properties of A_3B, perovskite, $MMo6X_8 \& M_2A_3X_3$ compounds were evaluated over crystal symmetries. In addition, research on the production methods and usage areas of superconductors are presented in this section.

Keywords

Superconductors, Crystallographic Structures, Type 1, Critical Temperature, Bardeen-Cooper-Schrieffer Theory

Contents

1.   Introduction

Materials that can conduct electrons between atoms without any resistance are defined as superconductors. When these materials reach the temperature at which superconductivity starts, they maintain their structure in a stable manner and do not show any heat, sound or other energy release [1–3]. This temperature at which superconductivity begins is defined as the critical temperature. However, some superconductors can exhibit these properties at very low temperature and energy conditions. Studies are generally carried out on alloys that show superconducting properties under high temperature conditions, rather than these materials. Because superconducting materials at very low temperatures is an inefficient and expensive process due to the high energy requirement [4–6].

There are generally two types of superconducting materials. It consists of basic materials used in places such as electrical transport systems from the first type of computer equipment [7,8]. The second type is the compounds formed by metallic structures, usually containing copper and lead [9].

Type 1 superconductors reach their critical temperature in atmospheric conditions (unpressurized conditions) in the temperature range of -273 °C to -265.35 °C. Some type 1 materials, on the other hand, require high pressure to reach their superconducting structure [10]. Sulfur, which is one of these materials, gains superconducting properties at -256.15 °C under pressure ($9.4 \times 10^{11} N/m^2$). Materials that can be given as examples of Type 1 superconductors are mercury, tin, aluminum and zinc [11–13].

Almost 50% of the elements in the periodic table can show superconducting properties. These materials are summarized in Figure 1.

The critical temperature value required by Type 2 materials to gain superconductivity is considerably higher than Type 1. In Type 2 compounds, under standard pressure, the highest critical temperature value is -138 °C ($HgBa_2Ca_2Cu_3O_8$). Another difference of Type 2 superconductors from Type 1 is that they are affected when exposed to a magnetic field [14,15].

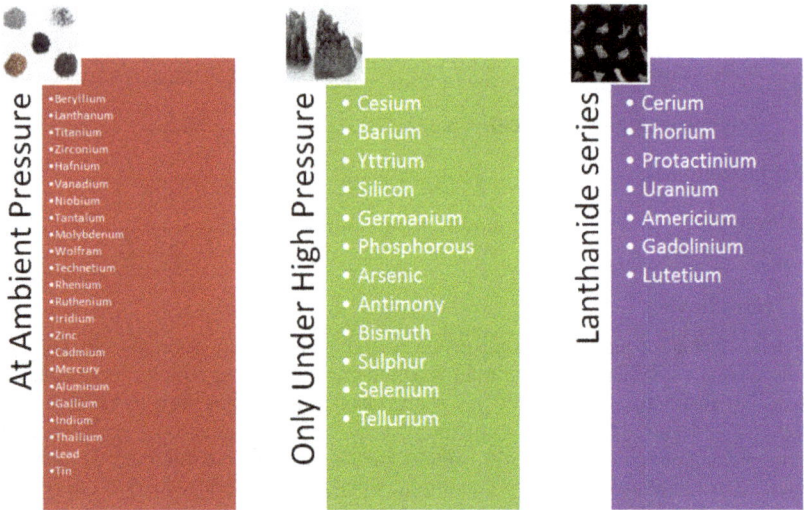

Figure 1.   Superconductor materials, (red) At ambient Pressure (green) Under High Pressure (purple) Lanthanides

In 1911, Heike Kamerlingh Onnes determined that when liquid helium was heated to -269°C, the electrical resistance of the material was disappeared. Thus, superconductivity is a phenomenon first understood, approaching close to 4 degrees Celsius of absolute temperature in Kelvin [16]. Twenty-two years after this work, German researchers Walther Meissner and Robert Ochsenfeld, while studying the behavior of matter at extremely cold temperatures, found that a superconducting material affects the magnetic field [17–19]. As in the working principle of the electric generator, a magnet moved in a conductor induces an electric current. This is specifically called the Meissner effect [20]. This effect is strong enough to levitate a superconducting material. Subsequent chronological developments in superconductors are shown in Figure 2.

2.   Type 1 Superconductors

This type of superconductors called as soft superconductors. Soft superconducting materials are also conductive at room temperature [21]. However, this value cannot be defined as superconducting [22]. These types of materials are generally composed of metals and metalloids [23–25]. These materials have a low critical temperature compared to type 2 [26,27]. Critical temperature values vary between -273°C and -263°C [28–32]. When Type 1 materials reach this temperature, their resistance and magnetic field properties drop dramatically and they become superconducting [33,34]. According to the

BCS theory put forward by Bardeen, Cooper, and Schrieffer in 1957, the slowing down of molecular activity creates a favorable environment for the Cooper effect, so that dipoles can easily overcome molecular barriers and act as free electrons [35–42]. Among the metals, the highest conductivity belongs to Au, Ag and Cu, respectively, but these three elements are not superconducting [43,44]. Because face-centered cubic lattice structures do not allow the flow of free electrons. Here, it is an important parameter that the cage structures are tightly packed and that they do not lose their properties even at low temperatures [45–48]. On the other hand, some metals with cubic face-centered lattice structure such as Pb can gain superconducting properties [49,50]. This is due to the mechanical properties of Pb such as low Young's modulus. Figure 3 present critical temperatures of Type 1 super conductors. Type-I superconductors differ from type-II superconductors by Ginzburg-Landau constant, κ. Type-I superconductors have $\kappa < 1/\sqrt{2}$ while type-II superconductors have $\kappa > 1/\sqrt{2}$ relationship of Ginzburg-Landau constant. κ value of a superconductor depends on ratio of penetration depth (λ) to coherence length (ζ).

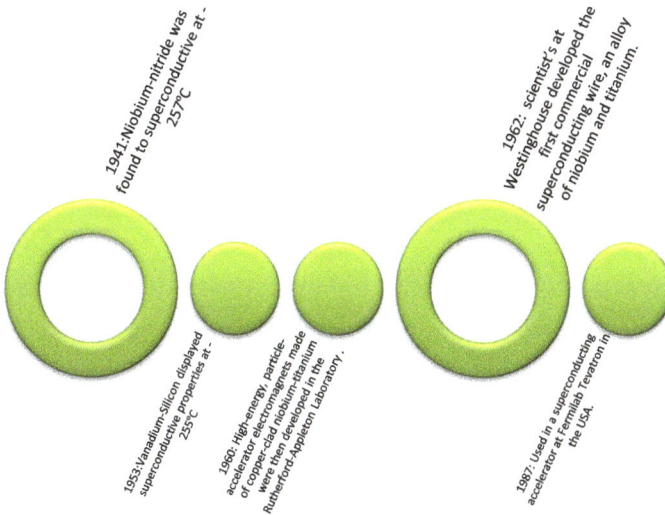

Figure 2. Chronological developments in superconductors.

Most of the elemental metals exhibit superconductivity at extremely low temperatures. Elemental metals are type-I superconductors with low critical transition temperatures, Tc. Type-I superconductors are generally metals and metalloids however, some compounds demonstrate type-I behavior by means of superconductivity. Intermetallic compounds are found to be type-I superconductors by various researchers.

TYPE 1 SUPERCONDUCTORS CRITICAL TEMPERATURES

Rhodium (Rh)	Lithium (Li)	Platinum (Pt)*	Tungsten (W)	Beryllium (Be)
Iridium (Ir)	Hafnium (Hf)	Uranium (U)	Titanium (Ti)	Ruthenium (Ru)
Cadmium (Cd)	Americium (Am)	Zirconium (Zr)	Osmium (Os)	Zinc (Zn)
Molybdenum (Mo)	Gallium (Ga)	Aluminum (Al)	Thorium (Th)	Protactinium (Pa)
Rhenium (Re)	Thallium (Tl)	Chromium (Cr)*	Palladium (Pd)*	Indium (In)
Tin (Sn)	Mercury (Hg)	Tantalum (Ta)	Lanthanum (La)	Lead (Pb)

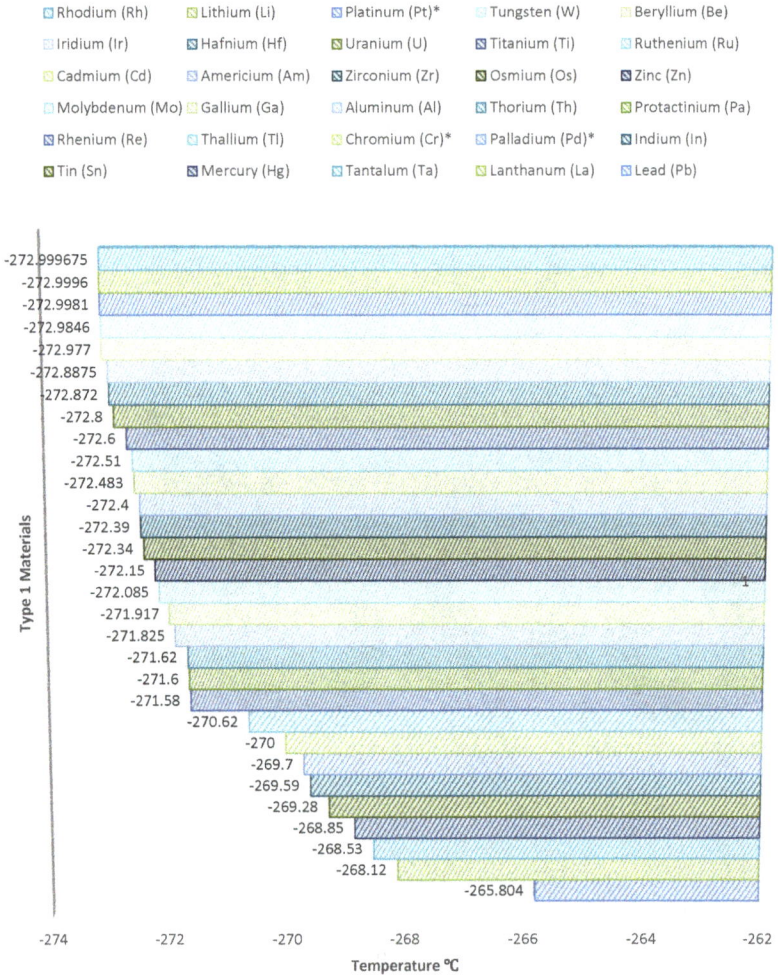

Figure 3. Critical Temperatures of Type 1 Superconductors.

3.   Structural properties of superconductors

The crystallographic structures of superconductors were evaluated on known compounds. Afterwards, it continued its development with the modification of these crystal groups. Understanding the crystal structure and properties of these materials will help find new superconductors in the future [51–55].

The Bardeen-Cooper-Schrieffer (BCS) theory, which was introduced in the 1950s, has been very useful for understanding the structural properties of superconductivity. According to the BCS theory, electron-phonon coupling pairs electrons with opposite spins and crystal momenta, resulting in a superconducting state. The cooper pair defined in superconductivity theory can move freely in lattice coordinates. Therefore, in BCS theory, electron pairs continue on their way in the alternatively created momentum space without encountering any resistance. As a result, it is not possible to talk about a stable lattice parameter for electron pairs according to BCS theory. While BCS theory does not give enough information about the synthesis of superconductors and what new materials will be, it helps to understand their superconducting mechanisms. From a chemical perspective, materials with known superconducting properties form phonons by attracting electrons with different wave vectors to each other in the lattice with the effect of lattice vibrations. These electron pairs can move through the lattice without encountering resistance. Δ is defined as the superconducting energy gap and an energy of 2Δ is required to break the electron pair in question. As can be understood from this, it is obvious that the energy from the atomic oscillation will be insufficient to break the cooper pair. Thus, it is seen that the electrons will not encounter any obstacles in the later stages of the process [56–59]. There is a relation between Δ and critical temperature as shown in equation 1.

$$2\Delta(0) = 3.52 k_B Tc \tag{1}$$

Here, $\Delta(0)$ denotes the theoretical superconducting energy gap at absolute temperature, while k_B denotes the Boltzmann constant. In BCS theory, the critical temperature is modified according to equation 2.

$$Tc = 1.14 \frac{h\omega D}{2\pi kB} e^{-1/N(EF)\lambda} \tag{2}$$

In Eq. 2 terms are defined as; E_F: Fermi level, k_B: Boltzmann constant, $N(E_F)$: Density of states, λ: Electron-phonon interaction parameter, h: Planck's constant and ωD: Debye frequency. According to Equation 2, the high value of all components other than the Boltzmann constant causes an increase in the critical temperature. Based on this, McMillan calculated the critical temperatures of classical soft superconducting materials, and the findings are shown in Table 1.

Superconductors - Materials and Applications
Materials Research Foundations 132 (2022) 146-165

Materials Research Forum LLC
https://doi.org/10.21741/9781644902110-8

Table 1 Critical Temperature of Superconductors calculated via Eq. 2.

Superconductor Materials	Tc (°C)
Nb	-251.15
Pb	-263.95
Nb_3Sn	-245.15
V_3Si	-233.15

Of course, these values may vary with the effect of pressure, the calculations are valid for ambient conditions.

In the phenomenon of superconductivity, starting from known superconducting compounds, an approach has been developed that materials in similar crystal lattice systems can easily undergo modifications and have this feature. For this reason, it will be beneficial for future studies to explain the structures of compounds, including Type 1 materials, over crystal lattices with superconducting properties [60–62].

4. A_3B structure superconductors

One of the superconductor compounds type is A_3B with P m3n space group. Figure 4 shows crystallographic space groups of A_3B compounds.

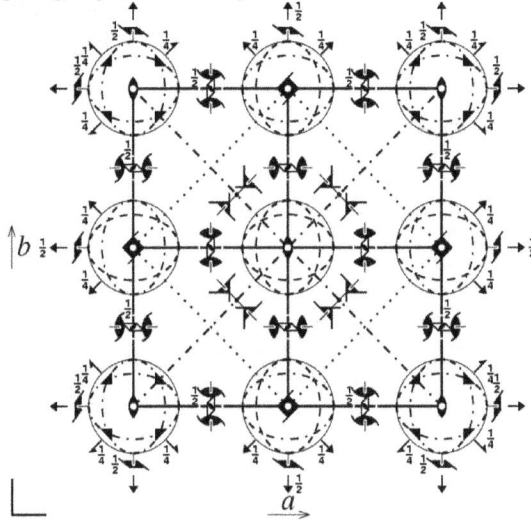

Figure 4. P m3n space group.

Cr_3Si and Nb_3Sn are examples of such superconducting compounds. While the B structure "Si, Sn" forms the center and corner atoms seen in Figure 4, there are A structures "Cr, Si" at 1/4-1/2 symmetry points along the x,y,z axis. According to Matthias' empirical rule, the critical temperature peaks when the valence electron/atom number ratio reaches approximately 4.7-6.4. This ratio was found to be 4.75 for Nb_3Ge, which is in A_3B structure, and the critical temperature value was determined as - 249.95°C [63–65].

5. MMo_6X_8 & $M_2A_3X_3$ structures superconductors

Superconductors in this structure consist of a combination of Metal + Molybdenum + Framework ($M+Mo_6+X_8$). In this type of superconductors, which can be arranged in triclinic, rhombohedral and hexagonal lattice systems, Mo can form compounds with (Te, Se and S) materials. $LaMo_6S_8$ is an example of superconductors of this nature. The MMo_6X_8 structure is located in the hexagonal lattice structure according to the P $6_3/m$ space group.

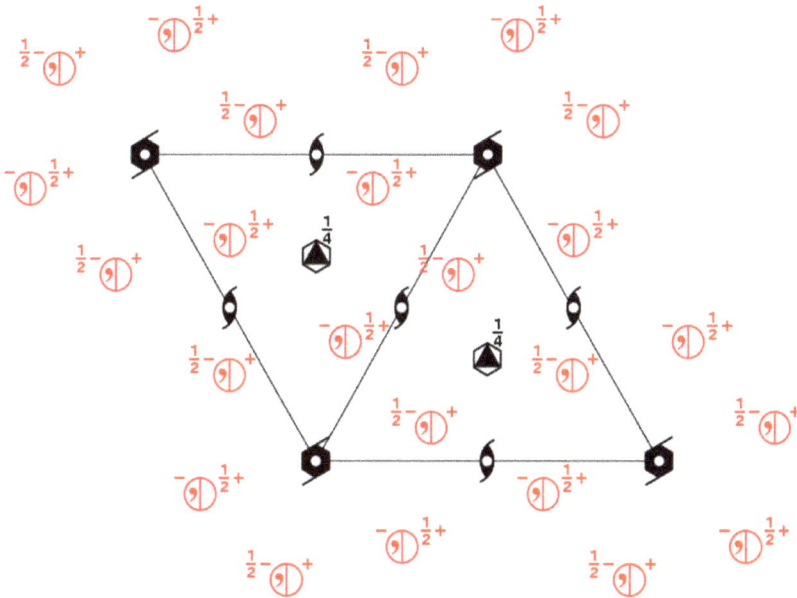

Figure 5. P $6_3/m$ space group.

In Figure 5, molybdenum atoms are framed by 8 X atoms located at (+,-1/2) coordinates. The X_8 frame is stacked forming a prism similar to the primitive cubic structure. Here, the distances between Mo_6 atoms on the lattice are not equal and different crystal orders may occur due to the formation of different X atoms. Therefore, new superconductors with the P 6m2 (Figure 6) space group such as $K_2Cr_3As_3$ have emerged.

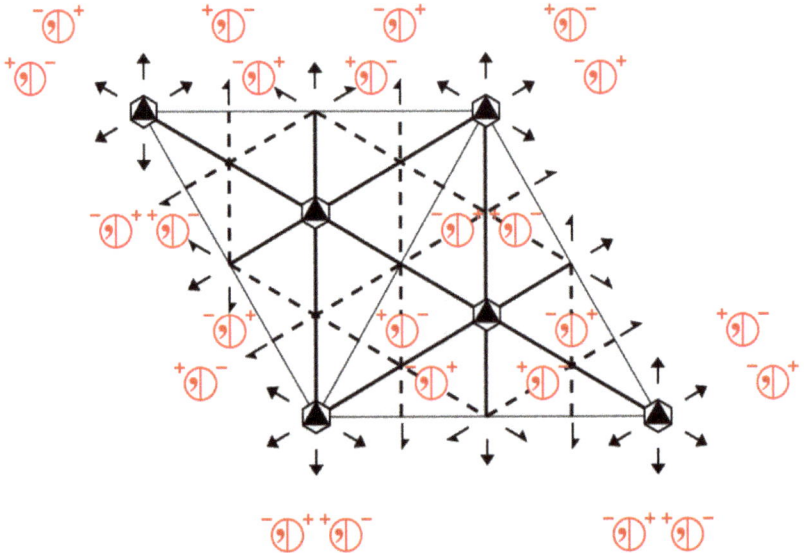

Figure 6. P 6m2 space group.

Here, along the c-axis of the hexagonal structure, semi-one-dimensional (Q-D1) $[Cr_3As_3]$ chains running their way. K atoms are located between these chains, 3 Cr atoms are arranged gradually in the ab plane, exhibiting similar behavior in arsenic [66]. As a result, hexagonal structures lined up along the c-axis seen (red ones) in Figure 6 are formed [67]. This causes a hexagonal structure similar to Mo_6X_8.

6. Cuprate superconductors structures

Perovskite structures are a common type of compound that works today with many positive properties as well as superconductivity. In these materials with ABO_3 composition, A generally consists of alkaline and rare earth elements, while B consists of transition metals. One of the most important perovskite structures with superconducting properties is $BaBiO_3$ and its two derivatives. These two derivatives are $BaPb_{1-x}Bi_xO_3$ and $Ba_{1-x}K_xBiO_3$, respectively, and their critical temperatures were determined as -260.15°C&-243.15°C. The $BaBiO_3$ structure, which has a cubic structure at high temperatures, transforms into a rhombohedral form, then a monoclinic structure, and finally a orthorhombic lattice system as the temperature decreases. This causes a high decrease in the amount of symmetry formed. Although charge imbalance occurs in the structure as a result of the transformations that occur with temperature change, Bi atoms are in stable crystallographic positions. Accordingly, the $Bi@O_6$ octahedra are in an alternative position within the lattice. The P $2_1/n$ symmetry formed at low temperatures is shown in Figure 7.

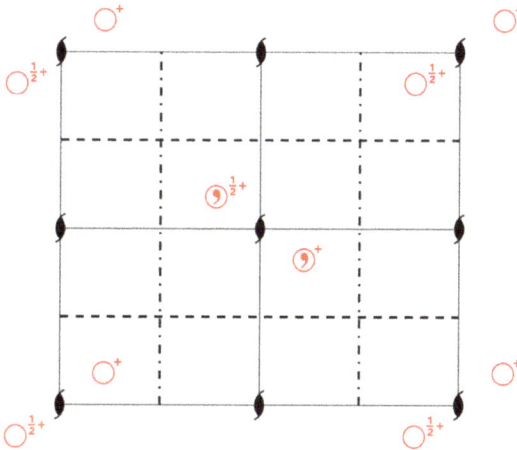

Figure 7. P $2_1/n$ space group

As can be seen in Figure 7, there are not many locations where atoms can be located, but the $Bi@O_3$ structure can be located at the coordinates in the center.

7.   Production of superconductors

Superconductor production is one of the most challenging topics in superconductor technology. Production of superconductors needs pure materials due to effect of impurities on superconducting properties. Crystal defects in material may change superconducting properties.

Superconductor wire production has vital place in superconducting applications. Grain size of superconducting wires is crucial for superconducting properties Grain size control greatly depends on temperature of production process.

Thin film superconductor properties may change with film thickness of material. Thin film thickness effects the vortex structure of material during superconducting

Superconductors - Materials and Applications   Materials Research Forum LLC
Materials Research Foundations 132 (2022) 146-165   https://doi.org/10.21741/9781644902110-8

8.   Wire production

Superconductor wires have applications in electric current transfer in various industries. Some of the commercialized superconductor materials for wire production are MgB_2 [45,48,68], Nb_3Al [34], Nb-Ti [29], Nb_3Sn [29]. Wire production techniques varies for different materials.

Wire production of superconductor materials can be conducted in-situ or ex-situ conditions. In-situ wire production of superconductors consists of synthesizing of intermetallic compounds in wire. Intermetallic compound formers are placed in wire structure then heat treatment for synthesis is applied.

MgB_2 superconductor is produced via in-situ technique. Due to their brittle nature, intermetallic compounds are not designed for mechanical stress carrier in wire structure [45]. Mg rod is covered with boride powder and placed in a metallic material that provide mechanical strength for wire. Following the completion of wire design heat treatment applied to produce MgB_2 compound [45].

Nb_3Al intermetallic compound is produced via in-situ technique. High stability under large electromagnetic force and relatively good mechanical properties of Nb_3Al are making this material an important superconductor wire material [34]. In-situ preparation of Nb_3Al consists of stochiometric Nb – Al mix is prepared via mechanical alloying [34]. Following the raw material preparation Rapid Heating/Quenching (RHQ) process is applied to wire [69]. RHQ process ensures that grain growth of material is limited and homogenous magnetic properties [34]. RHQ process increases the material temperature to synthesize intermetallic material then quenches to stop formation of unwanted phases and grain growth [70]. Heating of material is done by electrical current, supplied by Cu wheel which rolls over the wire, afterwards material is quenched in Ga bath to finish the RHQ [70].

Fully commercialized Nb_3Sn wires are produced and used in accelerator research projects [71]. Two techniques are used in Nb_3Sn wire production; Powder in Tube (PIT) and Internal Tin (IT) [71]. PIT technique consists of placing powders of Nb and Sn into a Cu tube and heated to produce Nb_3Sn phase in Cu matrix [29]. IT production of Nb_3Sn wires depends on Sn diffusion from bronze to Nb to obtain Nb_3Sn phase [29]. Doping of Ti to Nb_3Sn wires is found to be effective on improving magnetic properties [72].

General view of produced wires is given in Figure 8. Figure 8 shows that superconducting material is in the core of wire to perform superconducting properties and main structure of wire is the matrix that provide structural integrity and mechanical strength. Matrix of wire can play role in production of superconducting phase or may be only used for mechanical strength.

Superconductors - Materials and Applications
Materials Research Foundations 132 (2022) 146-165

Materials Research Forum LLC
https://doi.org/10.21741/9781644902110-8

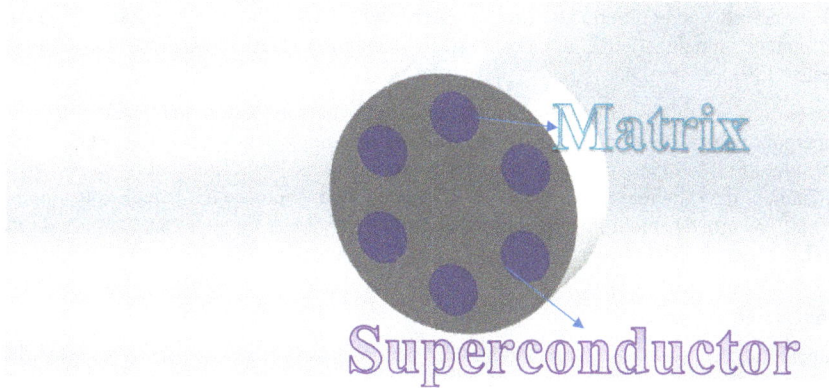

Figure 8. Schematic Illustration of Superconductor Wire.

9.   Thin films production

Superconducting thin film properties are depending on thin film thickness [70] and substrate properties [73]. The main substrate property by means of superconductivity is thermal conductivity [73]. Thermal conductivity has vital effect on superconductivity hence it directly effects the temperature distribution on thin film [73]. Thin film superconductivity based on film thickness and grain boundaries [74]. Grain boundaries in thin film are found to be deleterious for superconductivity since insulating grain boundaries suppress the current transfer [74].

Thin film thickness can determine the type of superconductivity [70]. Sn is known for exhibit superconductivity of Type-I or soft superconductor, Sn thin films of 50 and 100 nm are found to be Type-II superconductors [70].

Thin film production techniques are applicable for superconductor materials. Magnetron sputtering [75], Radio-Frequency sputtering [73], photolithography [70], electron beam evaporation [75], cold spray [76], vacuum plasma spraying [77] are some of the techniques which are used in superconductor thin film claddings. Techniques for thin film production may be adapted to produce superconducting films. Superconducting films can be used in particles accelerators and power devices [78].

10.   Superconductor applications

Superconductor technology has many application areas in engineering applications. Superconductors can be used in wire or thin film form. Suggested uses for superconducting materials include medical magnetic-imaging devices, magnetic energy-storage systems, motors, generators, transformers, computer parts, and very sensitive devices for measuring magnetic fields, voltages, or currents. The main advantages of devices made from superconductors are low power dissipation, high-speed operation, and high sensitivity. In

Superconductors - Materials and Applications   Materials Research Forum LLC
Materials Research Foundations 132 (2022) 146-165   https://doi.org/10.21741/9781644902110-8

addition, main superconductor use areas are cables for electrical transport, Magnetic Resonance Imaging (MRI) devices, high-field magnets, particle accelerators, induction heaters [29].

Developing of high temperature superconductors has increased application areas of superconductors. New technologies for rotating machines with electrical power is under consideration for superconducting material applications [79]. Eliminating of electrical resistance may increase the machine efficiency and decrease electrical consumption [79,80]. Magnetic shielding applications is another candidate for superconductor materials [79].

Conclusion

Superconductivity is a phenomenon that occurs when materials show no electrical resistance. In addition, in superconducting materials, there is a situation that prevents the penetration of the magnetic field.

The superconducting property can generally be achieved at very low temperatures. The main focus of the researchers on the subject is to obtain these properties at higher temperatures. While conventionally fabricated superconductors are called soft superconducting material (Type 1), those that gain this property at higher temperatures are named (Type 2). In this study, both the properties of soft superconductors and the properties of Type 2 superconductors obtained from these materials were investigated. Accordingly, it has been investigated whether the atoms allow electron flow according to the coordinates in the crystal lattice, and the relationship between superconducting feature formation and crystal structure has been determined in detail. The superconducting properties of structures such as $BaPb_{1-x}Bi_xO_3$ and $Ba_{1-x}K_xBiO_3$, which have different crystal lattice arrangements at varying temperatures, obtained as a result of polymorphic transformation were examined.

Researchers studying superconducting materials have focused on developing materials that will meet the need for energy transport and storage at higher temperatures. When these features are achieved, many popular applications such as super-fast trains, cables for electrical transport, high-field magnets, particle accelerators, induction heaters, Magnetic Resonance Imaging (MRI) devices, which are not affected by magnetic conditions in a frictionless environment, have been realized.

In this detailed research, some existing well-known superconductors, A_3B Structure Superconductors, MMo_6X_8 & $M_2A_3X_3$ Structures Superconductors, Cuprate Superconductors Structures, high pressure superconductors and interface superconductivity were investigated. The production methods of superconductors are emphasized, and wire and thin film applications are examined. In addition, by comparing the crystal structures, electronic structures and physical properties of different types of superconductors, the opinion that superconductors will one day be obtained at ambient pressure and room temperature has emerged strongly.

References

[1] Z. Li, L. Sang, P. Liu, Z. Yue, M.S. Fuhrer, Q. Xue, X. Wang, Atomically Thin Superconductors, 1904788 (2021) 1-15. https://doi.org/10.1002/smll.201904788

[2] G. Saito, Y. Yoshida, Organic Superconductors, 11 (2011) 124-145. https://doi.org/10.1002/tcr.201000039

[3] S. Sun, K. Liu, H. Lei, Type-I superconductivity in KBi 2 single crystals, (2016). https://doi.org/10.1088/0953-8984/28/8/085701

[4] C. Tarantini, W.L. Starch, N. Paudel, P.J. Lee, D.C. Larbalestier, Balachandran, 1644779 (2021) 1-8.

[5] Penetration depth study of the type-I superconductor PdTe 2, (2018).

[6] A. Vagov, Universal flux patterns and their interchange in superconductors between types I and II, Commun. Phys. (n.d.).

[7] S. Hameed, D. Pelc, Z.W. Anderson, A. Klein, R.J. Spieker, L. Yue, B. Das, J. Ramberger, M. Lukas, Y. Liu, M.J. Krogstad, R. Osborn, Y. Li, C. Leighton, R.M. Fernandes, M. Greven, Enhanced superconductivity and ferroelectric quantum criticality in plastically deformed strontium titanate, 21 (2022). https://doi.org/10.1038/s41563-021-01102-3

[8] B. Rosenstein, B.Y. Shapiro, PHYSICAL REVIEW B 100 , 054514 (2019) High-temperature superconductivity in single unit cell layer FeSe due to soft phonons in the interface layer of the SrTiO 3 substrate, 054514 (2019). https://doi.org/10.1103/PhysRevB.100.054514

[9] B.P. Dolan, C. Nash, J. Murugan, R. Adams, C. Kalousios, M. Spradlin, A. Volovich, Type I non-abelian superconductors in supersymmetric gauge theories, 11 (2007). https://doi.org/10.1088/1126-6708/2007/11/090

[10] T. Laboratories, M. Hill, Hard Superconductivity : Theory of the Motion of Abrikosov Flux Lines, (1963).

[11] H.T. Langhammer, T. Walther, Giant vortex states in type I superconductors simulated by Ginzburg - Landau equations, (n.d.).

[12] A.Y. Ganin, Y. Takabayashi, Y.Z. Khimyak, S. Margadonna, A. Tamai, M.J. Rosseinsky, K. Prassides, Bulk superconductivity at 38 K in a molecular system, (2008) 367-371. https://doi.org/10.1038/nmat2179

[13] J. Kitagawa, S. Hamamoto, N. Ishizu, from the Perspective of Materials Research, (2020).

[14] G. Shaw, S.B. Alvarez, J. Brisbois, L. Burger, W.A. Ortiz, B. Vanderheyden, A. V Silhanek, Magnetic Recording of Superconducting States, (2019) 1-17. https://doi.org/10.3390/met9101022

[15] P. Romano, A. Polcari, C. Cirillo, Drag Voltages in a Superconductor / Insulator / Ferromagnet Trilayer, (2021) 1-8. https://doi.org/10.3390/ma14247575

[16] M.I. Valerio-cuadros, D. Araujo, D. Chaves, F. Colauto, A. Augusta, M. De Oliveira, A. Marcos, H. De Andrade, T.H. Johansen, W.A. Ortiz, M. Motta, Superconducting Properties and Electron Scattering Mechanisms in a Nb Film with a Single Weak-Link Excavated by Focused Ion Beam, (2021). https://doi.org/10.3390/ma14237274

[17] B. Douine, K. Berger, Characterization of High-Temperature Superconductor Bulks for Electrical Machine Application magnetic, (2021). https://doi.org/10.3390/ma14071636

[18] W.A. Cooper, D. Brunetti, G. Tei, M. Nakatani, Vortices in a wedge made of a type-I superconductor Vortices in a wedge made of a type-I superconductor, (2015).

[19] J. Geng, J.M. Brooks, C.W. Bumby, R.A. Badcock, Time-varying magnetic field induced electric field across a superconducting loop : beyond dynamic, (2018).

[20] H. Matsuhata, C. Lee, K. Kihou, H. Eisaki, by electron microscopy, (2008) 1-2.

[21] J.R. Clem, S. Y, Flux-Line-Cutting Threshold in Type II Superconductors *, 39 (1980). https://doi.org/10.1007/BF00118073

[22] C.J. Boulter, J.O. Indekeu, Interfacial Tension and Interface Delocalization Phase, 19 (1998) 857-865. https://doi.org/10.1023/A:1022695023687

[23] J.G. Bednorz, K.A. Miiller, Condensed Matt Possible High T c Superconductivity in the Ba - L a - C u - 0 System, 193 (1986) 189-193. https://doi.org/10.1007/BF01303701

[24] X. Xu, M.D. Sumption, X. Peng, Internally Oxidized Nb 3 Sn Strands with Fine Grain Size and High Critical Current Density, (2015) 1346-1350. https://doi.org/10.1002/adma.201404335

[25] M. Santosh, Modeling of Critical Current Density of Bulk High T c Superconductors, 126 (2014) 808-810. https://doi.org/10.12693/APhysPolA.126.808

[26] R. Hott, R. Kleiner, T. Wolf, G. Zwicknagl, Review on Superconducting Materials, (1933) 1-59. https://doi.org/10.1002/3527600434.eap790

[27] R. Hott, R. Kleiner, T. Wolf, G. Zwicknagl, Review on Superconducting Materials Review on Superconducting Materials, 2016. https://doi.org/10.1002/3527600434.eap790

[28] Y. Zhang, X. Xu, Original Contributions, 112 (2021) 2-9.

[29] C. Yao, Y. Ma, iScience ll Superconducting materials : Challenges and opportunities for large-scale applications, ISCIENCE. 24 (2021) 102541. https://doi.org/10.1016/j.isci.2021.102541

[30] G. Gao, L. Wang, M. Li, J. Zhang, R.T. Howie, E. Gregoryanz, V. V Struzhkin, L. Wang, J.S. Tse, Superconducting binary hydrides : Theoretical predictions and experimental progresses, Mater. Today Phys. 21 (2021) 100546. https://doi.org/10.1016/j.mtphys.2021.100546

[31] A. Zhang, W. Jiang, X. Chen, X. Zhang, W. Lu, F. Chen, Z. Feng, S. Cao, J. Zhang, J. Ge, Anomalous magnetization jumps in granular Pb superconducting films, Curr. Appl. Phys. 35 (2022) 32-37. https://doi.org/10.1016/j.cap.2021.11.010

[32] Y. Artzi, Y. Yishay, M. Fanciulli, M. Jbara, A. Blank, Superconducting micro-resonators for electron spin resonance - the good , the bad , and the future, J. Magn. Reson. 334 (2022) 107102. https://doi.org/10.1016/j.jmr.2021.107102

[33] H. Fallah-arani, A. Sedghi, S. Baghshahi, R.S. Moakhar, N. Riahi-noori, N.J. Nodoushan, Bi-2223 superconductor ceramics added with cubic-shaped TiO2 nanoparticles_ Structural, microstructural, magnetic, and vortex pinning studies, J. Alloys Compd. 900 (2022) 163201. https://doi.org/10.1016/j.jallcom.2021.163201

[34] C. Yang, X. Yu, L. Liu, Z. Yu, Y. Chen, Y. Zhang, X. Pan, G. Yan, Y. Feng, Y. Zhao, Superconducting property improvement of RHQT Nb 3 Al wires through doping of Ti, J. Alloys Compd. 832 (2020) 154561. https://doi.org/10.1016/j.jallcom.2020.154561

[35] V.I. Kuznetsov, O. V Trofimov, Critical temperatures and critical currents of wide and narrow quasi-one-dimensional superconducting aluminum structures in zero magnetic field, Phys. C Supercond. Its Appl. (2022) 1354030. https://doi.org/10.1016/j.physc.2022.1354030

[36] R. Idczak, W. Nowak, M. Babij, V.H. Tran, Type-II superconductivity in cold-rolled tantalum, Phys. Lett. A. 384 (2020) 126750. https://doi.org/10.1016/j.physleta.2020.126750

[37] R. Idczak, W. Nowak, M. Babij, V.H. Tran, Physica C : Superconductivity and its applications Influence of severe plastic deformation on superconducting properties of Re and In, Phys. C Supercond. Its Appl. 590 (2021) 1353945. https://doi.org/10.1016/j.physc.2021.1353945

[38] C. Guo, H. Wang, X. Cai, W. Luo, Z. Huang, Y. Zhang, Q. Feng, Z. Gan, Physica C : Superconductivity and its applications High performance superconducting joint for MgB 2 films, Phys. C Supercond. Its Appl. 584 (2021) 1353863. https://doi.org/10.1016/j.physc.2021.1353863

[39] K. Van Bockstal, M. Slodička, Journal of Computational and Applied Error estimates for the full discretization of a nonlocal parabolic model for type-I superconductors, J. Comput. Appl. Math. 275 (2015) 516-526. https://doi.org/10.1016/j.cam.2014.01.022

[40] I.G. De Oliveira, L.R. Cadorim, A.R.D.C. Romaguera, E. Sardella, R.R. Gomes, M.M. Doria, The spike state in type-I mesoscopic superconductor, Phys. Lett. A. 406 (2021) 127457. https://doi.org/10.1016/j.physleta.2021.127457

[41] C.M.A. Lopes, M.I. Felisberti, Composite of low-density polyethylene and aluminum obtained from the recycling of postconsumer aseptic packaging, J. Appl. Polym. Sci. 101 (2006) 3183-3191. https://doi.org/10.1002/app.23406

[42] I.G. De Oliveira, The threshold temperature where type-I and type-II interchange in mesoscopic superconductors at the Bogomolnyi limit, Phys. Lett. A. 381 (2017) 1248-1254.. https://doi.org/10.1016/j.physleta.2017.01.032

[43] R. Folk, D. V Shopova, D.I. Uzunov, Fluctuation induced first order phase transition in thin films of type I superconductors, 281 (2001) 197-202. https://doi.org/10.1016/S0375-9601(01)00126-8

[44] V. Ya, R. Ya, ScienceDirect Solar sail with superconducting circular current-carrying wire, Adv. Sp. Res. 69 (2022) 664-676. https://doi.org/10.1016/j.asr.2021.10.052

[45] M. Balog, A. Rosova, B. Szundiova, L. Orovcik, P. Krizik, P. Svec, M. Kulich, L. Kopera, P. Kovac, I. Husek, A. Mohamed, H. Ibrahim, HITEMAL-an outer sheath material for MgB 2 superconductor wires : The effect of annealing at 595 - 655 ° C on the microstructure and properties, Mater. Des. 157 (2018) 12-23. https://doi.org/10.1016/j.matdes.2018.07.033

[46] N. Ishikawa, Defect production and recovery in high- T c superconductors irradiated with electrons and ions at low temperature, 263 (1998) 1924-1928. https://doi.org/10.1016/S0022-3115(98)00224-4

[47] S. Kumar, A.S. Dhavale, N.M. Chavan, S. Acharya, Superconducting niobium coating deposited using cold spray, Mater. Lett. 312 (2022) 131715. https://doi.org/10.1016/j.matlet.2022.131715

[48] X. Zou, W. Zhang, Q. Wang, L. Zheng, X. Yu, Z. Yu, H. Zhang, Y. Zhao, Preparation of MgB 2 superconducting wires by the rapid heating and quenching method, Mater. Lett. 244 (2019) 111-114. https://doi.org/10.1016/j.matlet.2019.02.067

[49] K.U. Leuven, R. Leiden, Wetting, prewetting and surface transitions in type-I superconductors, 251 (1995) 290-306. https://doi.org/10.1016/0921-4534(95)00421-1

[50] P. Wanderer, M.D. Anerella, A.F. Greene, E. Kelly, E. Willen, CQa transfer for industrial production of superconducting, (1995).

[51] R. Riedinger, A. Wallucks, I. Marinković, C. Löschnauer, M. Aspelmeyer, S. Hong, S. Gröblacher, Remote quantum entanglement between two micromechanical oscillators, Nature. 556 (2018) 473-477. https://doi.org/10.1038/s41586-018-0036-z

[52] N.M. Linke, D. Maslov, M. Roetteler, S. Debnath, C. Figgatt, K.A. Landsman, K. Wright, C. Monroe, Experimental comparison of two quantum computing

Materials Research Forum LLC
https://doi.org/10.21741/9781644902110-8

architectures, Proc. Natl. Acad. Sci. U. S. A. 114 (2017) 3305-3310. https://doi.org/10.1073/pnas.1618020114

[53] R. Barends, A. Shabani, L. Lamata, J. Kelly, A. Mezzacapo, U. Las Heras, R. Babbush, A.G. Fowler, B. Campbell, Y. Chen, Z. Chen, B. Chiaro, A. Dunsworth, E. Jeffrey, E. Lucero, A. Megrant, J.Y. Mutus, M. Neeley, C. Neill, P.J.J. O'Malley, C. Quintana, P. Roushan, D. Sank, A. Vainsencher, J. Wenner, T.C. White, E. Solano, H. Neven, J.M. Martinis, Digitized adiabatic quantum computing with a superconducting circuit, Nature. 534 (2016) 222-226. https://doi.org/10.1038/nature17658

[54] J.M. Gambetta, J.M. Chow, M. Steffen, Building logical qubits in a superconducting quantum computing system, Npj Quantum Inf. 3 (2017) 0-1. https://doi.org/10.1038/s41534-016-0004-0

[55] L. Dicarlo, M.D. Reed, L. Sun, B.R. Johnson, J.M. Chow, J.M. Gambetta, L. Frunzio, S.M. Girvin, M.H. Devoret, R.J. Schoelkopf, Preparation and measurement of three-qubit entanglement in a superconducting circuit, Nature. 467 (2010) 574-578. https://doi.org/10.1038/nature09416

[56] J.R. Schrieffer, Theory of super conductivity, Theory Supercond. (2018) 1-332. https://doi.org/10.1201/9780429495700

[57] J.F. Clark, F.J. Pinski, D.D. Johnson, P.A. Sterne, J.B. Staunton, B. Ginatempo, van Hove singularity induced L11 ordering in CuPt, Phys. Rev. Lett. 74 (1995) 3225-3228. https://doi.org/10.1103/PhysRevLett.74.3225

[58] J. Appel, Transition temperature of d-f-band superconductors, Phys. Rev. B. 8 (1973) 1079-1087. https://doi.org/10.1103/PhysRevB.8.1079

[59] P.B. Allen, Fermi-surface harmonics: A general method for nonspherical problems. Application to Boltzmann and Eliashberg equations, Phys. Rev. B. 13 (1976) 1416-1427. https://doi.org/10.1103/PhysRevB.13.1416

[60] L.R. Testardi, Structural instability and superconductivity in A-15 compounds, Rev. Mod. Phys. 47 (1975) 637-648. https://doi.org/10.1103/RevModPhys.47.637

[61] A. Ślebarski, P. Zajdel, M. Fijałkowski, M.M. Maśka, P. Witas, J. Goraus, Y. Fang, D.C. Arnold, M.B. Maple, The effective increase in atomic scale disorder by doping and superconductivity in Ca3Rh4Sn13, New J. Phys. 20 (2018). https://doi.org/10.1088/1367-2630/aae4a8

[62] H. Hayamizu, N. Kase, J. Akimitsu, Superconducting properties of Ca3T4Sn13 (T = Co, Rh, and Ir), J. Phys. Soc. Japan. 80 (2011) 2010-2012. https://doi.org/10.1143/JPSJS.80SA.SA114

[63] A.W. Sleight, J.L. Gillson, P.E. Bierstedt, High-temp SC in the BaPbBiO3 system, Solid State Commun. 88 (1993) 841-842. https://doi.org/10.1016/0038-1098(93)90253-J

Superconductors - Materials and Applications
Materials Research Foundations 132 (2022) 146-165

Materials Research Forum LLC
https://doi.org/10.21741/9781644902110-8

[64] R.J. Cava, B. Batlogg, J.J. Krajewski, R. Farrow, L.W. Rupp, A.E. White, K. Short, W.F. Peck, T. Kometani, Superconductivity near 30 K without copper: The Ba0.6K 0.4BiO3 perovskite, Nature. 332 (1988) 814-816. https://doi.org/10.1038/332814a0

[65] J.S. Manser, J.A. Christians, P. V. Kamat, Intriguing Optoelectronic Properties of Metal Halide Perovskites, Chem. Rev. 116 (2016) 12956-13008. https://doi.org/10.1021/acs.chemrev.6b00136

[66] Fischer, A. Treyvaud, R. Chevrel, M. Sergent, Superconductivity in the $RExMo6S8$, Solid State Commun. 88 (1993) 867-870. https://doi.org/10.1016/0038-1098(93)90259-P

[67] R. Chevrel, M. Sergent, J. Prigent, Sur de nouvelles phases sulfurées ternaires du molybdène, J. Solid State Chem. 3 (1971) 515-519. https://doi.org/10.1016/0022-4596(71)90095-8

[68] X. Xiong, Q. Wang, F. Yang, J. Feng, C. Li, G. Yan, Physica C : Superconductivity and its applications Improved superconducting properties of multifilament internal Mg diffusion processed MgB 2 wires by rapid thermal processing, Phys. C Supercond. Its Appl. 580 (2021) 1353800. https://doi.org/10.1016/j.physc.2020.1353800

[69] K. Nakamura, T. Takao, T. Ikeda, T. Higuchi, K. Tagawa, G. Iwaki, A. Description, Nb 3 Al Wire Development for Future Accelerator Magnets, 16 (2006) 1204-1207. https://doi.org/10.1109/TASC.2006.871298

[70] W. Bang, T.D. Morrison, K.D.D. Rathnayaka, I.F. Lyuksyutov, D.G. Naugle, W. Teizer, Characterization of superconducting Sn thin films and their application to ferromagnet-superconductor hybrids, Thin Solid Films. 676 (2019) 138-143. https://doi.org/10.1016/j.tsf.2019.02.033

[71] S. Farinon, T. Boutboul, A. Devred, D. Leroy, L. Oberli, Nb3Sn wire layout optimization to reducen cabling degradation, IEEE Trans. Appl. Supercond. 18 (2008) 984-988. https://doi.org/10.1109/TASC.2008.922299

[72] D.R. Dietderich, A. Godeke, Nb3Sn research and development in the USA - Wires and cables, Cryogenics (Guildf). 48 (2008) 331-340. https://doi.org/10.1016/j.cryogenics.2008.05.004

[73] J.P. Wu, H. Sen Chu, Substrate effects on intrinsic thermal stability and quench recovery for thin-film superconductors, Cryogenics (Guildf). 36 (1996) 925-935. https://doi.org/10.1016/S0011-2275(96)00083-5

[74] N.A. Khan, M. Mumtaz, How grain-boundaries influence the intergranular critical current density of Cu 1-xTl xBa 2Ca 2 Cu 4O 12-δ superconductor thin films?, J. Low Temp. Phys. 151 (2008) 1221-1229. https://doi.org/10.1007/s10909-008-9801-y

[75] A. Andreone, C. Aruta, M. Iavarone, F. Palomba, M.L. Russo, Microwave properties of RE - Ni 2 B 2 C ž RE s Y , Er / superconducting thin films, Scan. Electron Microsc. (1999) 141-149. https://doi.org/10.1016/S0921-4534(99)00305-6

[76] G. Ciovati, G. Cheng, U. Pudasaini, R.A. Rimmer, Multi-metallic conduction cooled superconducting radio-frequency cavity with high thermal stability, Supercond. Sci. Technol. 33 (2020). https://doi.org/10.1088/1361-6668/ab8d98

[77] E. Lugscheider, T. Weber, Coating technology, Fresenius' Zeitschrift Für Anal. Chemie. 333 (1989) 293-298. https://doi.org/10.1007/BF00572307

[78] K. Makise, T. Kawaguti, B. Shinozaki, Superconductor-insulator transitions in quench-condensed Bi films on different underlayers, Phys. E Low-Dimensional Syst. Nanostructures. 39 (2007) 30-36. https://doi.org/10.1016/j.physe.2006.12.040

[79] J. H. Durrell1 , M. D. Ainslie , D. Zhou, P. Vanderbemden, T. Bradshaw, S. Speller, M. Filipenko, D. A. Cardwell, Bulk superconductors: a roadmap to applications, Supercond. Sci. Technol. 31 (2018) 103501 https://doi.org/10.1088/1361-6668/aad7ce

[80] M. Chen, L. Donzel, M. Lakner, W. Paul, High temperature superconductors for power applications, Journal of the European Ceramic Society 24 (2004) 1815-1822. https://doi.org/10.1016/S0955-2219(03)00443-6

Superconductors - Materials and Applications   Materials Research Forum LLC
Materials Research Foundations 132 (2022) 166-178   https://doi.org/10.21741/9781644902110-9

Chapter 9

Oxide Superconductors

Srijita Basumallick

Asutosh College, 92, S.P. Mukherjee Road, Kolkata700026, India

Abstract

Unlike conductors and semiconductors, super conducting materials are relatively rare. Again, these materials exhibit super conductivity only at low temperature. In this chapter, an attempt has been made to present a brief historical background of discovery of super conductors. This is followed by presentation of unusual magnetic properties of super conducting materials, type 1 and type 2 super conducting materials, theories and hypothesis proposed to explain these properties of super conducting materials. Crystal structure of super conducting materials have been discussed.

Keywords

Type 1 Superconductor, Type 2 Superconductor, Yttrium Barium Copper Oxide, Cooper Pair, Meissner Effect

Contents

Superconductors - Materials and Applications   Materials Research Forum LLC
Materials Research Foundations 132 (2022) 166-178   https://doi.org/10.21741/9781644902110-9

1. Background

Superconductivity was first discovered in 1911. It was Heike Kamerlingh Onnes a dutch physicist who first observed superconductivity in mercury at 4 K [1]. He was the one who first discovered liquid He. But until 1986 extremely low critical temperature of all materials made superconductivity inappropriate for any technical application. It was when Georg Bednorz with Karl Müller [2] discovered lanthanum barium copper oxide with critical temperature 35 k the first ever HTS (high temperature superconductor). Followed by this Chu et al. discovered Yttrium barium copper oxide as superconductor at 93 K which is much higher than liquid nitrogen temperature (77 K) [3]. Superconductivity can be of two types; type 1 and type 2 or soft and hard superconductors. Usually type 1 superconductors are made of pure elements whereas type 2 superconductors are amalgam type or mainly alloys or oxide superconductor with defect in crystal structures [4]. Examples of type 1 superconductors pure Alluminium (Tc = 1.2 K), Mercury (Tc = 2K) or pure Lead (Tc = 7.19 K) [5]. Whereas type 2 superconductors include alloys like Nb_2Sn [5] (Tc18.3 K), Nb_3Ge [6] (Tc = 37 K) or Nb_3Al as well as oxide superconductor like $EuBa_2Cu_3O_x$ (EuBCO) (Tc = 93K)(7) or $GdBa_2Cu_3O_{7-x}$ (GdBCO) (Tc = 91K) [6].

It has been shown that induced magnetic field change in a different manner for type 1 and type 2 superconductors as applied magnetic field varies.

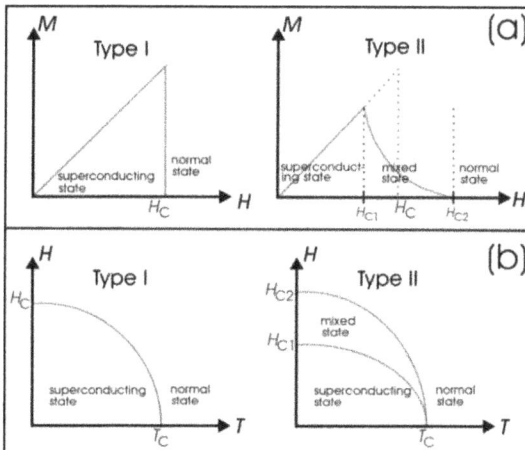

Figure 1: (a) Magnetisation M for type 1 and type 2 superconductors with change in applied magnetic field H. (b) Critical magnetic field vs temperature in kelvin for type 1 (showing single critical magnetic field below Tc) and type 2 (showing two critical magnetic field below Tc) [8].

Thus, type 1 superconductor has single critical magnetic field after which magnetisation vanishes where as type 2 superconductor has two critical fields. This critical magnetic field also depends on the temperature. Critical temperature shows vanishing of critical field indicating no more critical field or superconductivity above this (critical) point. Thus type 1 and type 2 superconductors bellow critical field and critical temperature shows properties of superconductor i.e. zero resistance. Several experiments and theories tried to explain this strange phenomenon. The difference between type1 and type 2 superconductors is tabulated in Table 1.

Table 1 Differences between type1 and type2 super conducting materials.

Type 1	Type2
These materials behave like perfect diamagnet.	They are not perfect diamagnet
There is only one critical magnetic field.	There are two critical magnetic fields.
They show total Meissner effect	They show partial Meissner effect.
They called soft super conductors	They called heard super conductors.

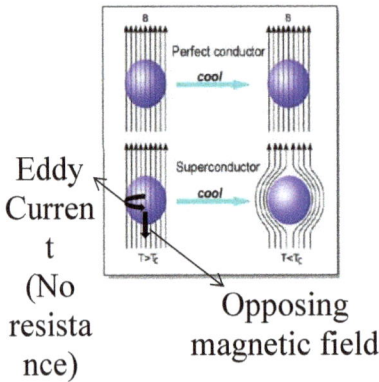

Figure 2: Schematic diagram of the Meissner effect(Left) [10]. Digital picture showing levitation of magnet on superconductor (Right) [11].

2.   Unusual properties super conducting materials and proposed theories and hypothesis

After Kamerlingh Onnes's discoverey of superconductivity in 1911. German physicists W. Meissner and R. Ochsenfeld discovered Meissner effect in 1933 [9]. In his simple experiment Meissner observed, if a superconductor was cooled bellow critical temperature it had help levitate a magnet. This may be explained simply as an eddy current formed inside the superconductor and goes on perpetually. For normal conductors the same eddy current will cause a repulsive force to an approaching magnet but dies off unless any one (magnet or conductor repulsing it) of them keep moving. Movement will induce change in flux density followed by constant repulsion. Finally, it (superconductor) keeps all magnetic field lines out making it's character like a perfect diamagnetic element.

To explain such effect several theoretical explanation came up. In 1935 London brothers; Fritz and Heinz London [12] showed that magnetic field (or flux denoted by B) inside superconductor is zero. Previously it was thought that due to zero internal resistance eddy's current does not change with fixed B. The main point is not only ($\partial B/\partial t$) but B is also zero. They pointed that it is not the disappearance of the resistance but rather the diamagnetism that took place in Messiner effect. London proposed super current (Js) is directly proportional to magnetic vector potential. They showed that

$$B_x = B_0 e^{\frac{-x}{\lambda}}$$

Where B_x is the flux at a depth of x distance from surface of superconductor. B_0 is flux outside and λ is penetration length

$$\lambda = \sqrt{\frac{m}{\mu_0 n_s e^2}}$$

n_s is number density (number/volume) of superconducting electrons, m is the mass of one electron or $9.1093837015 \times 10-31$ kg and e is charge of one electron or 1.6×10^{-19} C and μ_0 is permeability of free space. Also λ varies with temperature [13].

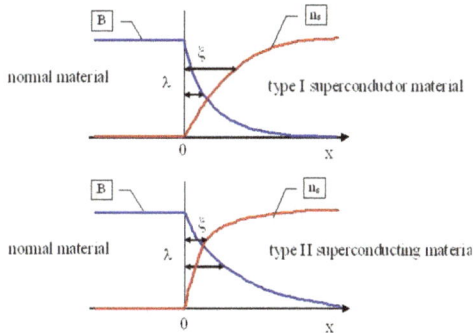

Figure 3: Penetration depth (blue line) and coherence length (red) vs distance (x) from surface of superconductor for Type 1 and Type 2 superconductors.

Superconductors - Materials and Applications   Materials Research Forum LLC
Materials Research Foundations 132 (2022) 166-178   https://doi.org/10.21741/9781644902110-9

But the London equation is based on macroscopic model, thus, cannot explain type 2 or oxide superconductor at relatively higher temperature. It is mainly valid near Tc and for homogeneous material.

Ginzburg and Landau two Russian physicist proposed in 1950 (V.L. Ginzburg and L.D. Landau, Zh. Eksp. Teor. Fiz. 20, 1064 (1950). English translation in: L. D. Landau), papers (Oxford: Pergamon Press, 1965) p. 546) [14] that came up with an advanced theory which deals with an inhomogeneous system. Their theory is based on calculation of free energy and as seen from Figure 1 (b). It is noted that the Hc value decreases continuously with T, thus, dHc/dT has no discontinuity showing it is a second order phase transition. Ginzburg and Landau mimicked normal ferromagnetic iron type system. Ginzburg and Landau uses helmoltz free energy arising for normal iron with an added term representing the kinetic energy of these particles or electron causing super current (considering quantum mechanical model of the system) in external magnetic field and external magnetic field energy density is represented by $\frac{H^2}{8\pi}$, H represent external magnetic field. From Ginzburg and Landau equation, one get another length scale coherence length ξ. That is in contrast to London's penetration length λ denote length up to perfect ordered (paired) electronic state. So, within the superconductor as we penetrate more inside surface the order develops and at the characteristic length of coherence length the order builds up fully and at characteristic length of penetration depth external flux repelled fully. Now from Ginzburg and Landau model, it is clearly understood that free energy decreases with penetration as ordering increases. Whereas, external magnetic field which is indicator of free and unpaired electron, decreases with penetration length. Now for type 2 super conductor there is partial penetration of magnetic flux in the field region between lower and upper critical field. Thus, in the region between Hc1 and Hc2 (Figure 1 (b)), superconductor and the normal region coexist. So, Ginzburg and Landau model can be reduced to free energy minimum model, where two phases are at equilibrium at the interface of normal and superconductor. Now k (Ginzburg and Landau parameter) = $\frac{\lambda}{\xi} > \frac{1}{\sqrt{2}}$ (indicate type 2 superconductor) and $\frac{\lambda}{\xi} < \frac{1}{\sqrt{2}}$ (indicate type 1 superconductor). They took complex order parameter (ψ) that depends on space in place of magnetization, where ψ^2 indicate number density of super conducting electrons in the superconductor. Though as such this theory is based on macroscopic model many year latter Gor'kov proved Ginzburg and Landau theorem is actually a limiting case of microscopic state. Details of derivation with a linearised Ginzburg and Landau equation give mathematical expression that converges to 1D harmonic oscillator and the maximum energy corresponding to the maximum allowed magnetic field gives upper critical field for type 2 super conductor given by Hc2 $=\frac{\hbar}{q\xi^2} = \frac{\varphi_0}{2\pi\xi^2}$ where $\frac{\hbar}{q} = \varphi_0 (flux\ quanta)$ and q= effective charge of super conducting electron. Thus, it leads to a new quantization of upper critical magnetic field. But this theory is also valid near Tc.

Superconductors - Materials and Applications
Materials Research Foundations 132 (2022) 166-178

Materials Research Forum LLC
https://doi.org/10.21741/9781644902110-9

3. Cooper pair model

Since pairing or ordering is the key idea of J. Bardeen, L. Cooper, and R. Schrieffer (BCS) [15], they proposed electron migration by formation of **cooper pair**. In BCS theory the exclusively derived considering wave nature of electron and energy of such two particles coupled with each other. Later, work of Ring. R. Doll and M. Näbauer [16] and Bascom S. Deaver, Jr. and William M. Fairbank [17] independently calculated value of $\varphi_0 (flux\ quanta)$ given by $\frac{h}{q} = \varphi_0 (flux\ quanta)$. It has been seen that flux quantum can give us the value of q effective charge of super electron. This value of q was found to be 2e. So, for superconductors there must be pair of electrons. Cooper proposed the pairing of electron happens through phonon. This is also supported by isotope effect. Since phonon means vibration of lattice which depends on mass. They found that the critical temperature depends on mass of the ions involved. But it can also happen through magnonexitron etc. Further Josephson proposed Josephson's junction and phonon mediated electric transfer. This mainly deals with the conception of tunnelling [18] through thin barrier (insulator plane) with a particular phase difference. Failure of phonon mediated electron transfer from isotope effect and spin fluctuation related d wave superconductor.

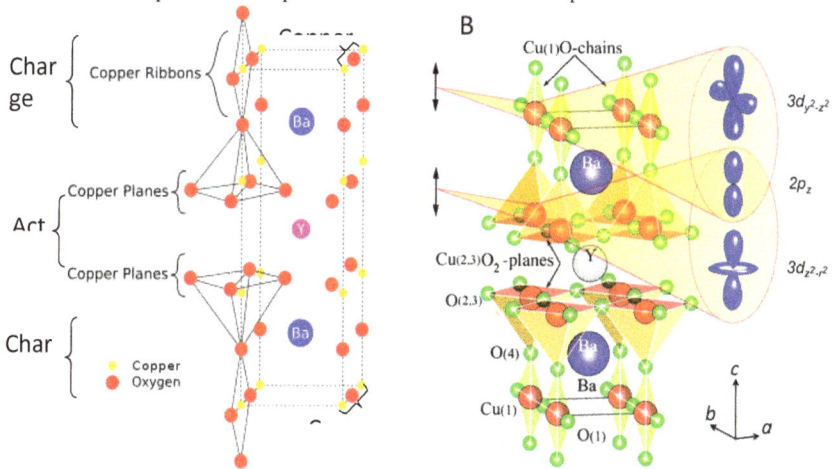

Figure 4: Crystal structure (part of the cell) of Yttrium barium copper oxide [19] (Left) Yttrium barium copper oxide crystal structure (part of the cell) with atomic orbital involved [20] (Right).

4. Crystal structure analysis of superconducting materials

Crystal structure of superconductor especially copper oxide based superconductor, ceramic superconductors are named using stoichiometric number. Bi-m2(n-1)n under BSCCO

Superconductors - Materials and Applications
Materials Research Foundations 132 (2022) 166-178

Materials Research Forum LLC
https://doi.org/10.21741/9781644902110-9

denote $Bi_mSr_2Ca_{n-1}Cu_nO_{2n+m+2}$ or Bi-2201 used to name $Bi_2Sr_2CuO_{6+\delta}$. These high temperature superconductor family is named Bi-HTS (high temperature superconductor). Here with increase in number of n critical temperature in increases. Thus Bi-2201 (34 K [21]), Bi-2212 (96 K [22]), Bi-2223 (110 K [23]), Bi-2234 (110 K [24]). In case of rare earth high temperature superconductor critical temperature is seen to reflect CuO_2 interplanar distance. This may be due to higher effective charge transfer in CuO_2 layers [25, 26]. Here are several common structure of HTS like TBCCO or HBCCO ($X_mBa_2Ca_{n-1}Cu_nO_{2n+m+2}$ or Tl/Hg-HTS or X-m2(n-1)n, where X denote Tl or Hg respectively) and RBCO ($YBa_2Cu_3O_{7-\delta}$; where Y can be replaced by Nd, Gd, Er, Yb and named accordingly known under 123-HTS). Apart from that Cu-HTS ($Cu_mBa_2Ca_{n-1}Cu_nO_{2n+m+2}$) is another important class of superconductor. When Ca in Cu-1212 i.e. $CuBa_2CaCu_2O_{7-\delta}$ is fully substituted by Y gives $YBa_2Cu_3O_{7-\delta}$ the exchange of Ca^{2+} with Y^{3+} creates extra doping. Yttrium barium copper oxide $YBCO(Y_1Ba_2Cu_3O_{7-\delta})$ (Figure 3 (a) and (b)) is the most studied superconductor material. It was also the first HTS with a critical temperature Tc (Tc = 92K) above the boiling point of liquid nitrogen [27]. If Y replaces other lanthanide with low radius (lanthanides have varying size because lower screening effect of d and f orbital leading to lanthanide contraction) will decrease interaction between two CuO_2 plane [24, 28, 29]. But precaution needs to be taken as rear earth metal tend to substitute Ba creating defect followed by loss in conductance [30]. Also, the inter planer coupling between two CuO2 conductor is further coupled across alkaline earth metal (Ba in YBCO) to form Josephson's coupling. Thus, Ba came into the picture as well as the CuO chain playing an important role in Josephson's coupling across the reservoir plane. The evidence of CuO_2 plane as charge carrier followed by CuO chains can be correlated through λ value at different direction 750 nm (along CuO2 plane) and 150 nm (perpendicular to CuO2 plane) in YBCO [31]. This clearly shows different concentration of super current carrier along CuO_2 plane and perpendicular to CuO_2 plane. Interestingly, substitution of oxygen by fluorine [32, 33] or chlorine [34] leads to increased conductance. In some HTS with increase in stoitiometry of Cu as well as Ca leads to increase in conductivity may be due to the CuO_2 conducting plane. There can be accumulation of interstitial oxygen when doping holes with electron this can create decrease in Tc.

For copper oxide semiconductor phase diagram indicating critical temperature of a substance versus hold doped clearly indicate antiferromagnetic behaviour of copper based ceramic materials at low hole doping concentration whereas metallic behaviour of the same at high hole doping concentration. But in intermediate region we can see superconductor being formed. Keeping hole doping concentration at stand if particular superconductor is heated produce pseudo metals which clearly doesn't have any clear cut conventional metallic or non metallic character. The effect of doping can be understood from figure 5. Here CuO_2 plane especially CuO chain plays key role in current conduction. This can be explained by visualizing copper dx^2-y^2 doesn't form an octahedral structure and contains hole. But this natural holes actually hinder conductivity because of Mott insulator type interaction. The hole trying to conduct has to jump from one site to other on a copper oxide plane where it has to seat with another natural hole. This makes it energetically unfavourable.

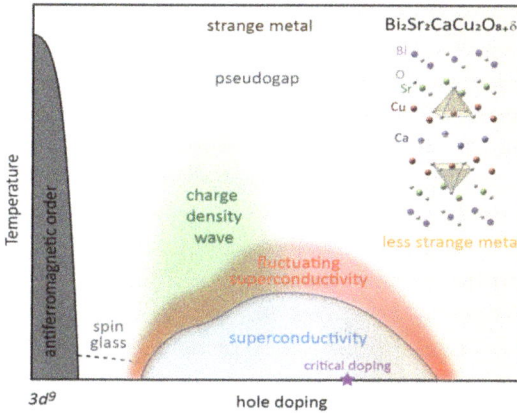

Figure 5: Phase diagram of cuprate superconductor ($Bi_2Sr_2CaCu_2O_{8+\delta}$) showing critical temperature vs hole-doping [35].

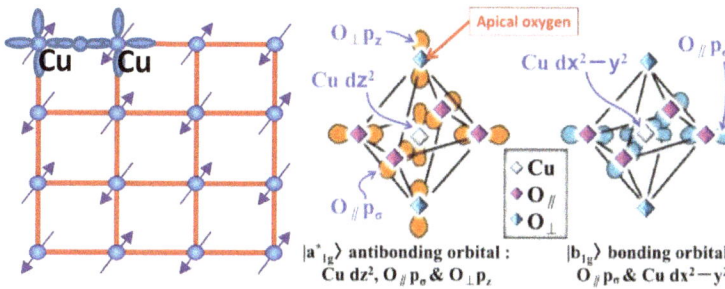

Figure 6: Schematic diagram showing anti ferromagnetic coupling (left) and MO of hole formation (right).

The undoped superconductor structure is antiferromagnetic insulator [36]. But Cu with d^9 electronic state has one unpaired electron and thus from band theory it seems to behave like a metal. After removal of 1 electron from typical d^9 copper to neighbouring copper, it becomes d8 with free dx^2-y^2 due to JT distortion. This free dx^2-y^2 is empty and creates in situ hole. But further electron transport will be hindered by hole repulsion. This will be

"Mott-Hubbard insulator" [37, 38]. Eventually it creates split into two hubbard bands with band gap "U" in conduction state. "t' being kinetic energy for such electronic shift in adjacent site. As kinetic energy (t) <<hubbard bands with band gap "U" it will stop delocalization of electron appear as an insulator. The super-exchange interaction (with an energy $J = t^2/U$) or virtual hopping creates antiferromagnetic arrangement as it favour antiparallel spins (follows Pauli exclusion principle) for better interaction. These high neel point insulating species conduct through defect as well as adding extra dope. In Cu based super conductor complication arises with dx^2-y^2 hybridization with pz of oxygen following change in lower hubberd band. In a recent work, it is seen existance of Cu (I), Cu (II) as well as Cu (III) in YBCO crystal which cause self doping [20].

5. Applications of oxide superconductor

It is really important to note that oxide superconductors have several applications like SQUID, MRI and maglev. To understand these mechanisms it is really important to know what is Josephson current. Josephson proposed that when two superconductors are joined through an insulator, where thickness of the insulator is typically around one nano meter, it is seen that current flows from one superconductor to the other. This can be explained by the tunneling effect of cooper pair electron through thin insulator barrier. Thus, under an external applied potential it is seen that super current starts flowing from one superconductor two the other through this thin film of insulator. As this super current doesn't have any resistance we will not find any current examined by a galvanometer kept in series with these two superconductors connected through an insulator. As a result potential drop across this superconductors junction will always be seen to be zero. The net current out of these two superconductors glued through insulator shows a phase equal to difference in phase of first superconductor current and second superconductor current. This is known as Josephsons DC current. Yet another type of Josephsons current is present which is known as AC Josephson's current. In case of AC Josephson's current when external applied potential reaches to a maxima normal electrons starts moving as well. Unlike cooper pairs all normal electrons do have resistance. As a result potential drop across the insulating junction start appearing. At the same time due to movement of normal electron from first superconductor to second superconductor the phase of first superconductor gets disturbed and the net phase of super current changes with change in external potential. This can be mathematically derived from some basic intuitive steps. If change in phase with time is denoted by w the angular frequency we can write $d\varphi/dt = w$ where φ is phase of current coming out from above-mentioned circuit. Again $2\pi v = w$, where v is frequency and w is angular velocity. Again, we know $E = hv$, h being planck constant or $E = \hbar*w$. Now writing energy in terms of potential we can write $E = 2qV$ where 2q is charge of one cooper pair and V is external potential applied. So, we can write $d\varphi/dt = 2qV/\hbar$ hence $\varphi = 2qVt/\hbar + \varphi_0$. This potential and time dependence of phase when replaced in original expression of current it shows a sinusoidal variation of current with potential variation. Thus, it gives an AC current of external applied field. This type of AC current is known as Josephsons AC current. Now in SQUID this type of superconductor insulator superconductor junction is farther coupled to form a ring with two junctions of insulators. As shown in figure 7 when current is passed across this set up current gets bifurcated in two superconductor insulators superconductor fragments. Now if a magnetic field is applied perpendicular to this spherical circuit there will be potential difference generated across booth superconductor insulators superconductor fragment. As a result there will be a net current coming out from this setup which is superimposition of two AC currents from two fragments of superconductor insulator superconductor parts of SQUID. This superimposed AC current frequency shows indication of tiniest magnetic field present under observation. SQUID magnetometer is known to detect very small magnetic field in medical as well as geological purposes. Namely SQUID is used in detection of magnetic pulses in neurone and other organ as well as it helps in detecting tiny amount of metal ore under earth crust without practically digging it out.

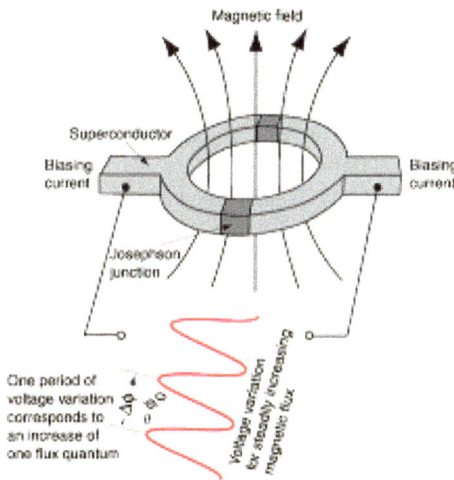

Figure 7: Schematic diagram of SQUID magnetometer.

Conclusions

In this chapter an attempt has been made to understand the properties of different types of super conducting materials and their broad classification as type 1 and type 2 super conducting materials. Needless to mention that oxide superconductors fall within type 2 superconductors. Genesis of super conductivity has been explained in terms of different existing theories and hypothesis. Elaborate discussions have been made on most accepted cooper pair. Some applications of superconductors in medical and geological field have been discussed.

Superconductors - Materials and Applications   Materials Research Forum LLC
Materials Research Foundations 132 (2022) 166-178   https://doi.org/10.21741/9781644902110-9

References

[1] Onnes, H. K., The resistance of pure mercury at helium temperatures, Commun. Phys. Lab. Univ. Leiden. 12 (1911) 120.

[2] Bednorz, J. G.; Müller, K. A., Possible highTc superconductivity in the Bf Lf C- O system, Zeitschrift für Physik B Condensed Matter. 64 (1986) 189-193. https://doi.org/10.1007/BF01303701

[3] Wu, M.K.; Ashburn, J.; Torng, C.; Hor, P.; Meng, R.; Gao, L.; Huang, Z.; Wang, Y. Q.; Chu, C., Superconductivity at 93 K in a New Mixed-Phase Y-Ba-Cu-O Compound System at Ambient Pressure, Physical review letters. 58 (1987) 908-910. https://doi.org/10.1103/PhysRevLett.58.908

[4] Humphreys, C. J., Ceramic Superconductors. In Concise Encyclopedia of Advanced Ceramic Materials, Brook, R. J., Ed. Pergamon: Oxford, 1991; pp 67-73. https://doi.org/10.1016/B978-0-08-034720-2.50028-9

[5] Eisenstein, J., Superconducting Elements, Reviews of Modern Physics. 26 (1954) 277-291. https://doi.org/10.1103/RevModPhys.26.277

[6] Oya, G.I.; Saur, E. J., Preparation of Nb3Ge films by chemical transport reaction and their critical properties, Journal of Low Temperature Physics. 34 (1979) 569-583. https://doi.org/10.1007/BF00114941

[7] Malavasi, L.; Tamburini, U. A.; Galinetto, P.; Ghigna, P.; Flor, G., The High-Temperature Superconductor EuBa2Cu3O6 + x: Role of Thermal History on Microstructure and Superconducting Properties, Journal of Materials Synthesis and Processing. 9 (2001) 31-37. https://doi.org/10.1023/A:1011334631235

[8] Santoso, I. k-space Microscopy of Bi2Sr2CaCu2O8+? :Fermiology and Many-body Effects. 2008.

[9] Meissner, W.; Ochsenfeld, R., Ein neuer Effekt bei Eintritt der Supraleitfähigkeit, Naturwissenschaften. 21 (1933) 787-788. https://doi.org/10.1007/BF01504252

[10] Meissner effect. https://en.wikipedia.org/wiki/Meissner_effect (accessed 28th March 2022).

[11] Shiva Effect of magnetic field: Meissner Effect Explaination. https://semesters.in/effect-of-magnetic-field-explanation-of-meissner-effect-for-btech-bsc-engineering-physics/.

[12] London, F.; London, H.; Lindemann, F. A., The electromagnetic equations of the supraconductor, Proceedings of the Royal Society of London. Series A - Mathematical and Physical Sciences. 149 (1935) 71-88. https://doi.org/10.1098/rspa.1935.0048

[13] de Launay, J.; Steele, M. C., On Temperature Dependence of Penetration Depth in Superconductors, Physical Review. 73 (1948) 1450-1454. https://doi.org/10.1103/PhysRev.73.1450

[14] 73 - ON THE THEORY OF SUPERCONDUCTIVITY. In Collected Papers of L.D. Landau, Ter Haar, D., Ed. Pergamon: 1965; pp 546-568. https://doi.org/10.1016/B978-0-08-010586-4.50078-X

[15] Bardeen, J.; Cooper, L. N.; Schrieffer, J. R., Theory of Superconductivity, Physical Review. 108 (1957) 1175-1204. https://doi.org/10.1103/PhysRev.108.1175

[16] Doll, R.; Näbauer, M., Experimental Proof of Magnetic Flux Quantization in a Superconducting Ring, Physical Review Letters. 7 (1961) 51-52. https://doi.org/10.1103/PhysRevLett.7.51

[17] Deaver, B. S.; Fairbank, W. M., Experimental Evidence for Quantized Flux in Superconducting Cylinders, Physical Review Letters. 7 (1961) 43-46. https://doi.org/10.1103/PhysRevLett.7.43

[18] Josephson, B. D., The discovery of tunnelling supercurrents, Reviews of Modern Physics. 46 (1974) 251-254. https://doi.org/10.1103/RevModPhys.46.251

[19] Yttrium barium copper oxide. https://en.wikipedia.org/wiki/Yttrium_barium_copper_oxide.

[20] Magnuson, M.; Schmitt, T.; Strocov, V. N.; Schlappa, J.; Kalabukhov, A. S.; Duda, L. C., Self-doping processes between planes and chains in the metal-to-superconductor transition of YBa2Cu3O6.9, Scientific Reports. 4 (2014) 7017. https://doi.org/10.1038/srep07017

[21] Feng, D. L.; Damascelli, A.; Shen, K. M.; Motoyama, N.; Lu, D. H.; Eisaki, H.; Shimizu, K.; Shimoyama Ji, J. I.; Kishio, K.; Kaneko, N.; Greven, M.; Gu, G. D.; Zhou, X. J.; Kim, C.; Ronning, F.; Armitage, N. P.; Shen, Z. X., Electronic structure of the trilayer cuprate superconductor Bi(2)Sr(2)Ca(2)Cu(3)O(10+delta), Phys Rev Lett. 88 (2002) 107001.

[22] Eisaki, H.; Kaneko, N.; Feng, D. L.; Damascelli, A.; Mang, P. K.; Shen, K. M.; Shen, Z. X.; Greven, M., Effect of chemical inhomogeneity in bismuth-based copper oxide superconductors, Physical Review B. 69 (2004) 064512. https://doi.org/10.1103/PhysRevB.69.064512

[23] Chu, C. W., High-temperature superconducting materials: a decade of impressive advancement of T/sub c, IEEE Transactions on Applied Superconductivity. 7 (1997) 80-89. https://doi.org/10.1109/77.614424

[24] Lösch, S.; Budin, H.; Eibl, O.; Hartmann, M.; Rentschler, T.; Rygula, M.; Kemmler-Sack, S.; Huebener, R. P., Structural analysis and superconducting properties of Pb-free Bi2Sr2Ca3Cu4Oy (n=4, nominal composition), Physica C: Superconductivity. 177 (1991) 271-280. https://doi.org/10.1016/0921-4534(91)90479-I

[25] Lin JG, Huang CY, Xue YY, Chu CW, Cao XW, Ho JC. Origin of the R-ion effect on RBa2Cu3O7, Physical Review B. 51 (1995) 12900-12903. https://doi.org/10.1103/PhysRevB.51.12900

[26] Tallon, J. L., Oxygen in High-Tc Cuprate Superconductors. In Frontiers in Superconducting Materials, Narlikar, A. V., Ed. Springer Berlin Heidelberg: Berlin, Heidelberg, 2005; pp 295-330. https://doi.org/10.1007/3-540-27294-1_7

[27] Zoller, P.; Glaser, J.; Ehmann, A.; Schulz, C.; Wischert, W.; Kemmler-Sack, S.; Nissel, T.; Huebener, R. P., Superconductivity and processing of single Bi−O layered

Superconductors - Materials and Applications   Materials Research Forum LLC
Materials Research Foundations 132 (2022) 166-178   https://doi.org/10.21741/9781644902110-9

cuprates in the (Bi, Pb)−Sr−(Ca, Y)−Cu−O system, Zeitschrift für Physik B Condensed Matter. 96 (1995) 505-509. https://doi.org/10.1007/BF01313848

[28] Yamauchi, H.; Tamura, T.; Wu, X.-J.; Adachi, S.; Tanaka, S., A Missing Link is Found: a Novel Homologous Series of Superconducting Pb-Based Cuprates, Japanese Journal of Applied Physics. 34 (1995) L349-L351. https://doi.org/10.1143/JJAP.34.L349

[29] Tamura, T.; Adachi, S.; Wu, X. J.; Tatsuki, T.; Tanabe, K., Pb-1223 cuprate superconductor with Tc above 120 K synthesized under high pressure, Physica C: Superconductivity. 277 (1997) 1-6. https://doi.org/10.1016/S0921-4534(97)00052-X

[30] Attfield, J. P.; Kharlanov, A. L.; McAllister, J. A., Cation effects in doped La2CuO4 superconductors, Nature. 394 (1998) 157-159. https://doi.org/10.1038/28120

[31] Kirtley, J. R.; Tsuei, C. C.; Moler, K. A., Temperature Dependence of the Half-Integer Magnetic Flux Quantum, Science (New York, N.Y.). 285 (1999) 1373-1375. https://doi.org/10.1126/science.285.5432.1373

[32] Karppinen, M.; Fukuoka, A.; Niinistö, L.; Yamauchi, H., Determination of oxygen content and metal valences in oxide superconductors by chemical methods, Superconductor Science and Technology. 9 (1996) 121-135. https://doi.org/10.1088/0953-2048/9/3/001

[33] Sadewasser S, Schilling JS, Hermann AM. Pressure-dependent oxygen diffusion in superconducting Tl2Ba2CuO6+δ, YBa2Cu3O7-δ, and HgBa2CuO4+δ: Measurement and model calculation, Physical Review B. 62 (2000) 9155-9162.

[34] Ohta, Y.; Tohyama, T.; Maekawa, S., Apex oxygen and critical temperature in copper oxide superconductors: Universal correlation with the stability of local singlets, Physical Review B. 43 (1991) 2968-2982. https://doi.org/10.1103/PhysRevB.43.2968

[35] Shen, P. Z.-X. Cuprate Superconductors. https://arpes.stanford.edu/research/quantum-materials/cuprate-superconductors.

[36] Takagi H, Ido T, Ishibashi S, Uota M, Uchida S, Tokura Y.Superconductor-to-nonsuperconductor transition in (La(1-x)Srx)2CuO4 as investigated by transport and magnetic measurements., Physical Review B. 40 (1989) 2254-2261. https://doi.org/10.1103/PhysRevB.40.2254

[37] Hubbard, J.; Flowers, B. H., Electron correlations in narrow energy bands, Proceedings of the Royal Society of London. Series A. Mathematical and Physical Sciences. 276 (1963) 238-257. https://doi.org/10.1098/rspa.1963.0204

[38] Mott, N., The Basis of the Electron Theory of Metals, with Special Reference to the Transition Metals, Proceedings of the Physical Society. Section A. 62 (2002) 416. https://doi.org/10.1088/0370-1298/62/7/303

Superconductors - Materials and Applications   Materials Research Forum LLC
Materials Research Foundations 132 (2022) 179-193   https://doi.org/10.21741/9781644902110-10

Chapter 10

High Temperature Superconductors: Materials and Applications

M.S. Hasan[1] and S.S. Ali[2*]

[1] Department of Physics, The University of Lahore, Lahore 54000, Pakistan

[2] School of Physical Sciences, University of the Punjab, Lahore 54590, Pakistan

* shahbaz.sps@pu.edu.pk

Abstract

A ceramic high temperature superconductor called cuprate was discovered in 1986 by Bednorz and Muller at 135 K critical temperature (T_c). Ladik and Bierman demonstrated the idea of high temperature superconductivity (HTSC) by essential excitation of electrons in one chain of double standard DNA. The cuprates are consisted of stiff, fragile, and hard properties which are considered as the negative impact of cuprates superconductors. With the passage of time such drawback were removed and the cuprate superconductors were made capable for various applications. The breakthrough came in the history of superconductors when the iron based HTSC were discovered in 2006. High temperature H_2S superconductors are designed at high pressure and T_c greater than 200 K in 2015. From the time of discovery of HTSC, superconductivity is used in different types of fields such as medical, electrical, magnetic, optics, recording media, microwave and communication devices.

Keywords

HTSC, Critical Temperature, Types of HTSC, Energy Storage System, Applications of HTSC

Abbreviations used

USOs	Unidentified superconducting objects
RT	Room temperature
HTSC	High temperature superconductors
T_c	Critical temperature
TM	Transition metals
AFM	Antiferromagnetic
RE	Rare earth
REBCO	Rare earth elements doped barium-copper oxide
H	Magnetic field
J_c	Critical current density

LHD	Large helical device
FFHR	Force free helical reactor
ESS	Energy storage system
SMES	Superconducting magnetic energy storage
LTSC	Low temperature superconductors
MW	Mega watt
PM	Permanent magnets
MRI	Magnetic resonance imaging

Contents

1. Introduction

Unrevealed superconductivity having critical temperature around room temperature (RT) was described during the 1970 – 80. More normally were elaborated after the innovation of cuprates. The discovery of mercury as a superconductive material, various other metals were determined showing superconductive nature at ultra-low temperatures. Niobium is the metal which has critical temperature of 9.25 K. A15 superconductors were designed in 1950 demonstrating maximum 23.2 K critical temperature [1]. A ceramic high temperature superconductor called cuprates was discovered in 1986 by Bednorz and Muller at 135 K critical temperature [2] and won the Nobel Prize. The list of different low temperature superconductors along with critical temperature is given in Table 1. The phenomena is becoming more popular with the reports of transition temperature far above the RT as higher pressures investigation about H_2S [3]. Al – C – Al (sandwich structure) is earlier

Superconductors - Materials and Applications   Materials Research Forum LLC
Materials Research Foundations 132 (2022) 179-193   https://doi.org/10.21741/9781644902110-10

example, where the stronger reliance on current due to external magnetic field is similar to Josephson DC tunneling current at 300 K [4]. Anilin Black (AB) presented a postulate for RT superconductivity and associated to Al – C – Al system, which constructed the high conductive quasi 1 – D domains [5]. Ladik and Bierman gave the idea of high temperature superconductivity by essential excitation of electrons in one chain of double standard DNA [6]. The small metallic crystal structures rooted on vapor deposit thin films are also part of the phenomena. Huge surface to volume ratio is obtained while coating the crystals with different dielectrics where the electrons in the neighborhood of surface have unlike characteristics as compare to bulk materials [7]. Inside the dielectric having frequencies higher than the Debye frequencies the polarization waves are excited. Therefore, the critical temperature for surface conductivity lies in the range of 102 – 104 K. CdF_2 originated sandwich structure consisted of boron inserted layers is the example of room temperature superconductors. For such type of superconductors at 319 K zero resistance was observed [8]. The diamagnetic susceptibility and resistivity were observed at 700 K transition temperature of carbon nanotubes [9].

Table 1 *Different types of superconductors along with superconducting transition temperature T_c[1].*

Type of superconductor	Compound	Transition temperature (K)
Simple metal superconductors	Al	1.17
	In	3.41
	Hg	4.20
	Sn	3.72
	Pb	7.20
	Nb	9.25
A15 superconductors	Nb_3Sn	18
	Nb_3Ge	23.2
Fullerene superconductors	$C_{60}Rb_3$	31
Cuprate superconductors	$La_{2-x}Sr_xCuO_4$	38
	$YBa_2Cu_3O_7$	93
	$Bi_2Sr_2Ca_2Cu_3O_{10}$	107
	$Tl_2Ba_2Ca_2Cu_3O_{10}$	125
	$HgBa_2Ca_2Cu_3O_8$	135
Magnesium diboride superconductors	MgB_2	39
Iron based superconductors	FeSe	39
	$LaO_{0.89}F_{0.11}FeAs$	8
	$Sr_{0.5}Sm_{0.5}FeAsF$	56

Superconductors - Materials and Applications   Materials Research Forum LLC
Materials Research Foundations 132 (2022) 179-193   https://doi.org/10.21741/9781644902110-10

The discovery of cuprates opened the novel age in research of superconductivity. In very small period of time high transition temperatures were obtained go beyond the values of liquid nitrogen for new features of superconductors. Although, the cuprates are consisted of brittle, easily fragile, and hard to handle properties so these are considered as the negative impact of cuprates superconductor's [1]. With the passage of time such drawback were removed and the cuprate superconductors were made capable for various applications. For longer period of time it was considered that cuprates have distinctive properties and are accurate high temperature superconductors. After this another HTSC MgB_2 superconductor was designed at 39 K critical temperature. Although, this T_c value is much lesser than the T_c of cuprates, however is much better than the conventional superconductors. The breakthrough came in the history of superconductors when the iron based HTSC were discovered in 2006 [10]. Utmost value of T_c for iron based HTSC is 56 K which is also less than the value of cuprates. Currently, HTSC based H_2S superconductors are designed at high pressure and T_c greater than 200 K in 2015. Such outcome is amazing because the results reveal that hydrogen originated compounds are favorite candidates to understand room temperature superconductors [3]. To attain the superconductivity beyond the limits the pressure is applied for possible applications. Actually, such compounds are consisted of layer materials. Alike phase diagrams are demonstrated by the iron based superconductors and cuprates and magnitude of T_c is investigated through the inserted charge carriers. Both types of superconductors act as a magnetic materials during the insulation [1].

2.   Science of HTSC

HTSC superconductors have various types of families. These superconductors have layered structure of Cu_2O which are distinguished by insulating layers. Insulating layers act as reservoir to provide the charge carriers to the layers of Cu_2O. Superconductivity also arises in the layers of Cu_2O. Cuprates were not discovered by chance, the stronger electron lattice interactions are required for the purpose of T_c is the base of cuprates. Some features are also discussed by BCS theory, but this theory has also some limitations like lattice frequency and number of free carriers. Therefore, the BCS theory is unable to explain the HTSC completely because of limitations on T_c [1]. Bednorz et al., observed that few superconductors have T_c value rather than particular low carrier density [2]. From this investigation they concluded that the pairing amongst lattice and carriers must be existed and it is not explained by BCS theory. In order to attain stronger pairing they considered that polaron structure may do work where the concept of Jahn Teller polaron is considered [11]. Polaron is type of quasi particle in which new entity is formed by the combination lattice deformation and carriers. Two of these quasi particles form together for superconductivity achieving which have distinction to the conventional cooper couple. Such behavior reduces the number of pairs inside the unit volume and has unlikely overlapping [1]. The exceptional stronger relations leading the superconducting T_c are due to the development of these quasi particles and their bound state is probable. Concurrently, the charge carrier density can be quite diminutive. Such critical subjects were considered in order to find the appropriate materials for superconductivity.

Bednorz and Muller took the initiative through $SrTiO_3$ (Perovskite) and investigated that such materials are suitable for superconductivity but the negative impact was low T_c [12]. The superconductivity could not be confirmed by the fabrication of new materials like nickelate and perovskite type oxide. Ultimately, they worked on HTSC cuprates. Hence, the perception of bipolaron process of HTSC was comprehend and continued. Though, after this development there is no existence of consensus still today after the application of various types of theories. In cuprates it is dependent on strength of carriers [13]. Except from the effect of isotope on T_c, various other novel effects of isotopes were also detected like energy gap and penetration depth.

3.   Nickel based HTSC

Nickel based high temperature superconductor materials were studied by Congcong Le *et al.*, and found that the compounds are consisted of layers with anti-perovskite structures. The compounds were manufactured by mixed anions nickel complexes. The structure of layers was designed in various other TM materials like $La_2B_2Se_2O_3$ where, B is Mn, Fe, Co. For such materials it was revealed that materials are consisted of host collinear AFM positions like the iron based HTSC. The Fermi energy was managed by two e_g d-orbital having entirely independent inplane kinematics. They also investigated that superconductivity of these materials is characterized by great competition amongst expanded s and d waves pairing symmetries [14].

Figure 1 (a) shows the promise to construct both e_g orbitals for powerful contribution in mixed anion nickel octahedral complex structures for inplane kinematics. The plan is to turn the complex and attach them such that apical oxygen may design the square lattice as given in Figure 1 (c). In the current scenario $(B_2M_2O_2)^{2-}$ having layer sheets is consisted of sharing tilt octahedral (Ni_2M_2O) and also nickel atom is enclosed by couple of axial oxygen ions and four M ions. Figure 1 (a) shows the two $d_{z^2}$and $d_{x^2-y^2}$ orbitals prior to the revolving are branded as $d_{x^2-y^2}$ and $d_{xz/yz}$ in innovative axis management. The $d_{x^2-y^2}$attains in-plane kinematics by oxygen and new $d_{xz/yz}$ makes strong pair with M ions and keeps in-plane kinematics via M ions. Due to the in-plane mirror regularity the two orbital are entirely decouple [14].

Such structure demonstrated in Figure 1 (a) and (c) is applied in $La_2Ni_2Se_2O_3$ layered materials as illustrated in Figure 1 (b), these are consisted of $(Ni_2Se_2O)^{2-}$. From this it was concluded that super exchange AFM pairings are exploited in nickel originated materials to create collinear AFM condition as similar to magnetic state original materials of iron based superconductor [15]. Two independent electronic band structures are formed by two e_g orbitals and are controlled by low energy electronic physics. Considered that super exchange interactions of AFM created the superconducting pairing. The powerful competition amongst d and extensive s waves pairing symmetries lead to characterize the compounds. Under the electron doping the d wave may develop into high competitive

Superconductors - Materials and Applications   Materials Research Forum LLC
Materials Research Foundations 132 (2022) 179-193   https://doi.org/10.21741/9781644902110-10

under the electron insertion or through the adjustment of lattice constants. On the other hand the s wave is supported via doping of holes. Such outcomes may give the phase diagram consisting of possible symmetry broken pairing states [14].

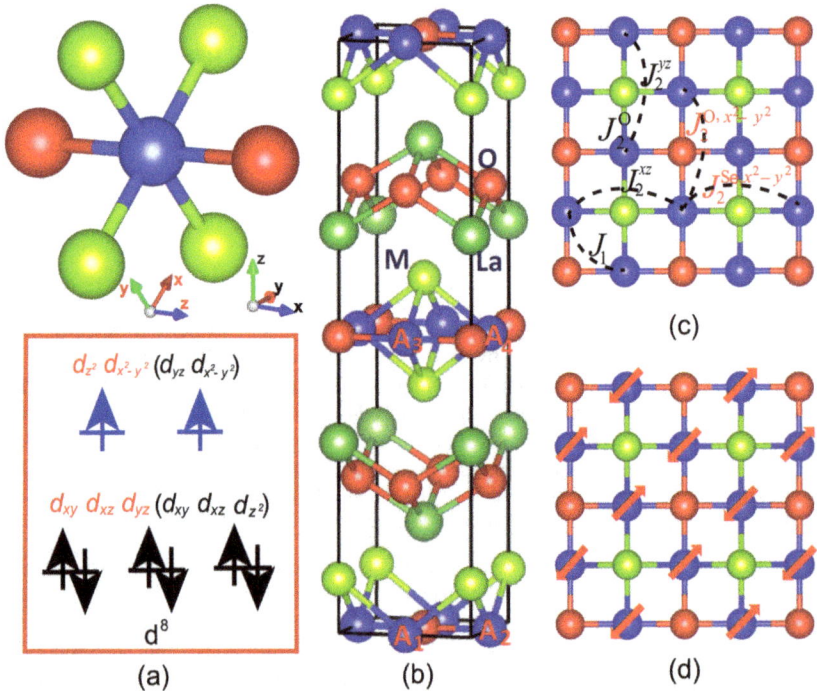

Figure 1 [14]: (Color online) 2-D layers structure, energy splitting of d-orbital, arrangement of electron occupation and magnetic configuration (a) BM_4O_2 octahedral system with blue is transition metal, green is chalcogens and red is oxygen. (b) $La_2B_2M_2O_3$. (c) B_2M_2O layer in ab plane. (d) Collinear AFM state of C-type.

4.   HTSC for fusion reactors

HTSC magnets are becoming potential candidates for future applications in the fields of commercial fusion reactors. For this purpose the remountable and demountable HTSC are considered for tokamak and helical reactors. Highly thermally stable and low cryogenic materials are favorites for such reactors [16]. HTSC and especially REBCO, where RE indicates the rare earth elements and BCO is barium copper oxides tapes have prominent aspects like greater cryogenic constancy alike the thermal ability of distinctive metals [17], low cryogenic potential at high working temperature [18], high Magnetic field (H) and high critical current density (J_c) [19]. HTSC magnets are normally applicable in fusion reactors in order to enhance the "H" to manage the high power fusion plasma where the large temperature margin is needed to produce huge nuclear heat [18]. Various supplementary attractive alternatives are existed for utilization of superior HTSC magnets accumulated through small coil or magnet sectors: demountable or remountable HTSC [20, 21] magnets and also joint-winding of such HTSC magnets [22, 23]. This fabrication procedure and magnet designs are also described as segmented fabrication and HTSC magnets, correspondingly. Remountable HTSC magnets are composed of magnet segments or coil segments with demountable combination. These designs suggest benefits like simplicity of fabrication of big and composite superconducting magnets, simple contact with internal structural constituents for instance diverters and blankets, and unproblematic substitution of unsuccessful magnet sectors. Joint-winding procedure is suggested particularly for LHD kind fusion reactors FFHR-d1 and c1, where FFHR is originally force free helical reactor [24, 25]. Here, the single pitch or half pitch HTSC helical conductors segments wound the helical coils by the application of permanent joints (not demountable). The advantage of such design is easy fabrication. While, the other designs need various resistive joints, planned procedure is acceptable due to high thermal capacities and low refrigeration energies of HTSC magnets [16].

5.   HTSC magnetic energy storage for power applications

Energetic and proficient characteristics are shown by the superconducting magnets whereas applied as an energy storage system (ESS) in speedy bidirectional transfer of electrical power with grid. Superconducting coil capacitors are required for diverse applications of ESS. The design of superconducting coil is expensive and has limitations such as the quench problems AC losses, magnetic field. The power electronic controllers and converters are obligatory for the utilization of superconducting coils as an energy storage device. Transmission lines constantly find issues such as inrush current, lightning surge current because of switching-on of electrical tools, bird short circuit and tree falling, which generate unsteadiness. [26-29]. The active and reactive power exchange capabilities of domestic structures, the dimensions or ESS connected to the power system must be enhanced [30-32]. Different ESS such as superconducting magnetic energy storage (SMES), flywheel energy storage, energy stored by batteries and supercapacitors are employed for permanence purposes because of their huge power absorption and transfer ability [33]. SMES is one of the most effectual and competent sources of energy storage. Characteristics of storage for various technologies are illustrated in Table 2. By the conversion of superconducting tapes from LTSC to HTSC technological devices, the efficiency of SMES was also observed to be enhanced. Although HTSC wire is quite costly and needed huge quantity for SMES coil, it is still price valuable as of power system constancy and low-cost liquid nitrogen refrigeration points of view. Hybrid SMES systems are designed by the combination of SMES and ESS devices [34-36]. This causes the increase in system's energy capacity and decline the capacity capital [37].

Table 2: *Various characteristics of storage technologies [37].*

Characteristics	Energy storage systems (ESS)			
	Flywheels	Super capacitors	SMES	Battery
Power rating (MW)	1.65	0.1	10 – 100	50
Response time (cycle)	1	¼	¼	¼
Discharge time (m)	2	1	1 – 30	1 – 480
Efficiency (%)	90	95	95	85
Lifetime	20	30	30	10
Maturity	Commercial	Commercial/ research	Research	Commercial

6.   HTSC materials based on bismuth

The superconducting compounds are considered as most attractive substances in the field of material science and engineering due to their unique characteristics. Also, have the properties of most capable manufacture compounds for different uses in numerous fields like electronics, instrumentation and power engineering etc. Currently, two such usages of HTSC primarily, low current electronics of such as nonlinear and passive superconducting elements, digital gadgets and SQUID electronics [38] and secondly high current – voltage apparatus such as stimulating devices and current – voltage controllers with the power of MW, transformers having 1.5 MW capacity, transmission line sectors for the power of 440 MW [39, 40]. In case of superconducting electrical gadgets, should be consisted of large critical currents and thermodynamic critical fields, large T_c and higher pinning force. Furthermore, the superconducting compounds should also have suitable mechanical properties e.g., strength, rigidity and stiffness. Contemporary, volumetric ceramic HTSC compounds are consisted of multiphase and inhomogeneous crystalline arrangements. The characteristics are able to demean ultimately. Structural materials are composed of film or single crystal HTSC for low-current electronics. Though, for electronic devices, nonlinear characteristics of superconductors are of immense technical and realistic attention for the purpose of transfer of currents through the volume. In order to alter the magnetic field the non-linearity in properties of current and voltage of HTSC were disclosed of higher harmonic constituents in signal reply. On the basis of such characteristics the materials are employed in magneto sensitive sensors and in various devices to generate and convert signals. Y-Ba-Cu-O and Bi-Sr-Ca-Cu-O are HTSC polycrystalline ceramic compounds consisted of small critical currents and characteristics of non-linearity. Such characteristics are fundamentally depended on micro and nanostructure. Such aspects lead to generate structural HTSC compounds with predefined nonlinear characteristics [40].

7. HTSC in co-axial magnetic gear

Gear is mechanical formation that conveys rotational motion among the parts of a machine. According to transformation of moment and speed in functions of revolving motion, the gear has capability to transport the motion between the parts. Suppose in the case of wind turbines, the wind is not enough to produce the electricity. Revolution motion of generator along with poles quantity are directly associated to frequency as well as voltage. The voltage and frequency attained through synchronous generators are required to be synchronized according to parameters of wind turbines. A gear system by means of special turning ratio must be under synchronization with the generator of wind turbine speed. Same condition is applied in gears of vehicles as, the mechanical engine has the constant speed but speed of medium can be increased or decreased. Hence, particular procedure of vehicle needs the gear kit to adjust the speed of wheel to speed of engine. The particular engineering works are done in order to control the mechanical pieces of any machine. However, for the control of different applications in the engines of vehicle the manual gear is not appropriate. These have numerous drawbacks because of resistance and vibration due to the mechanical interaction amongst the hooks of gear system. Unwanted situations for the gear materials like exhaustion, heat and rust may happen. Lubrication is one procedure to overcome such conditions, although it is not appropriate for all functions. Mechanical gear works noisily and is susceptible to overwork. Hence, direct drive mechanisms are recommended to get rid of present issues. As the direct drive procedures have no gears, so have high efficiency. Yet, direct drive mechanisms are not proper for every use. Magnetic gears are potential contestant to reduce all given issues faced by conventional gears. For example, magnetic gears have ability to convey rotational speed exclusive of any mechanical interactions and have no drawbacks in situation of overwork [41]. Different ways are there for the manufacturing of mechanical and magnetic gears. Normally, 50–200 kNm/m^3 are the torque density limitations of manual gear. Harmonic is category of magnetic gears with highest torque density 150 kNm/m^3 [42, 43]. Though torque densities of the mechanical gears are higher than the magnetic counterparts, the significance of magnetic gears inclined to amplify due to the unique characteristics. Also, magnetic gears belong to categories of renewable and sustainable energy resources. Moreover, the applications of renewable energy resources to reduce the environmental contamination formulate the magnetic gear environment friendly. The working principles of both magnetic and mechanical gears are same. The rule of function is probably similar in both systems. Figure 2 (a) and (b) illustrates the working mechanisms of both mechanical and magnetic gears, respectively [41].

The permanent type magnet was employed in the rotor of auto control gear with the rotation ratio of fifty and one. In 1972, N. Liang presented the auto control magnetic gear for vehicles [44]. Regarding cost point of view the liquid nitrogen for HTSC is one of the most prominent usages in cryogenic temperatures. In some conditions the cooling expense is not

Materials Research Forum LLC

https://doi.org/10.21741/9781644902110-10

of attention. Such as, usage of superconducting wire in MRI is most significant application of superconductivity regardless of the cooling shortcoming. With the advancement of cooling tools, the prospect of the potential usages of superconductivity will expand more. Besides of above mentioned applications of magnetic gears, limited losses of currents by means of HTSC are also significant features of auto gears [15, 45-47]. Various prominent and popular applications of auto magnetic gears based on HTSC are proposed to be major and foremost outcomes of permanent type magnets [48].

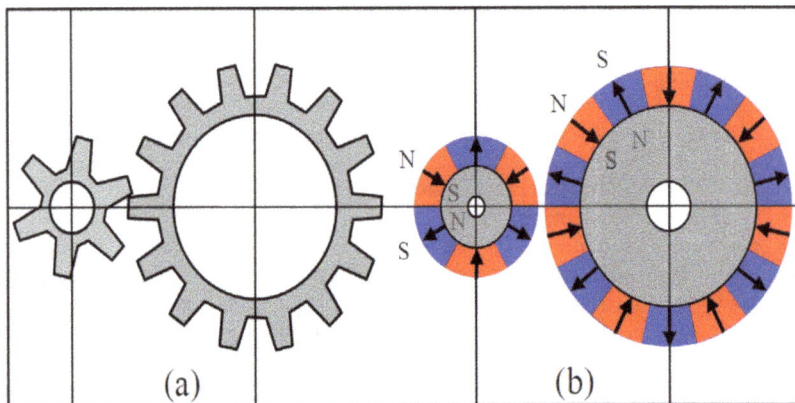

Figure 7.2 [41]: Schematic representation of (a) mechanical and (b) magnetic gears.

Conclusions

HTSCs also called cuprates were discovered in 1986 by Bednorz and Muller at 135 K critical temperature (T_c). The stronger reliance on current due to external magnetic field (H) is similar to Josephson DC tunneling current. HTSC superconductors have various types of families. CdF_2 based structures consisted of boron inserted layers showed zero resistance at 319 K. The discovery of cuprates opened the novel age in research of superconductivity. Cuprates have properties of brittleness, easily fragile, and hard to handle so these are considered as negative impacts of cuprates superconductors. Cuprate superconductors were made capable for various applications after the removal of such impacts. HT MgB_2 superconductors were designed at 39 K critical temperature. The breakthrough came in the history of superconductors when the iron based HTSC were discovered in 2006. H_2S based superconductors are designed at high pressure at T_c greater than 200 K in 2015. Nickel based high temperature superconductor materials are consisted of layers with anti-perovskite structures. HTSC magnets are becoming potential candidates for future applications in the fields of commercial fusion reactors. For this purpose the remountable and demountable HTSC are considered for tokamak and helical reactors. Due

Superconductors - Materials and Applications   Materials Research Forum LLC
Materials Research Foundations 132 (2022) 179-193   https://doi.org/10.21741/9781644902110-10

to the unique physical properties, superconducting compounds are assumed as one of most attractive substances in basic physics of condensed matter from time of detection because of novel physical characteristics.

References

[1] A. Bussmann-Holder, H. Keller, High-temperature superconductors: underlying physics and applications, Zeitschrift für Naturforschung B 75 (2020) 3-14. https://doi.org/10.1515/znb-2019-0103

[2] J.G. Bednorz, K.A. Müller, Possible high Tc superconductivity in the Ba− La− Cu− O system, Zeitschrift für Physik B Condensed Matter 64 (1986) 189-193. https://doi.org/10.1007/BF01303701

[3] A. Drozdov, M. Eremets, I. Troyan, V. Ksenofontov, S.I. Shylin, Conventional superconductivity at 203 kelvin at high pressures in the sulfur hydride system, Nature 525 (2015) 73-76. https://doi.org/10.1038/nature14964

[4] K. Antonowicz, Possible superconductivity at room temperature, Nature 247 (1974) 358-360. https://doi.org/10.1038/247358a0

[5] J. Langer, Unusual properties of the aniline black: does the superconductivity exist at room temperature?, Solid State Communications 26 (1978) 839-844. https://doi.org/10.1016/0038-1098(78)90755-X

[6] J. Ladik, A. Bierman, On the possibility of room-temperature superconductivity in double stranded DNA, Physics Letters A 29 (1969) 636-637. https://doi.org/10.1016/0375-9601(69)91140-2

[7] Y.S. Barash, V.L. Ginzburg, Some problems in the theory of van der Waals forces, Soviet Physics Uspekhi 27 (1984) 467. https://doi.org/10.1070/PU1984v027n07ABEH004025

[8] D. Eagles, Three small systems showing probable room-temperature superconductivity, Physica C: Superconductivity 483 (2012) 82-85. https://doi.org/10.1016/j.physc.2012.07.011

[9] N. Bagraev, E. Brilinskaya, E.Y. Danilovskii, L. Klyachkin, A. Malyarenko, V. Romanov, The de Haas-van Alphen effect in nanostructures of cadmium fluoride, Semiconductors 46 (2012) 87-92. https://doi.org/10.1134/S1063782612010022

[10] Y. Kamihara, H. Hiramatsu, M. Hirano, R. Kawamura, H. Yanagi, T. Kamiya, H. Hosono, Iron-based layered superconductor: LaOFeP, Journal of the American Chemical Society 128 (2006) 10012-10013. https://doi.org/10.1021/ja063355c

[11] H. KH, ock, H. Nickisch, H. Thomas, Helv. Phys. Acta 56 (1983) 237.

[12] H. Fröhlich, Theory of the superconducting state. I. The ground state at the absolute zero of temperature, Physical Review 79 (1950) 845. https://doi.org/10.1103/PhysRev.79.845

[13] J. Franck, Physical properties of high temperature superconductors IV, World Sci, Singapore (1994) 189. https://doi.org/10.1142/9789814440981_0004

[14] C. Le, J. Zeng, Y. Gu, G.-H. Cao, J. Hu, A possible family of Ni-based high temperature superconductors, Science Bulletin 63 (2018) 957-963. https://doi.org/10.1016/j.scib.2018.06.005

[15] P. Dai, Antiferromagnetic order and spin dynamics in iron-based superconductors, Reviews of Modern Physics 87 (2015) 855. https://doi.org/10.1103/RevModPhys.87.855

[16] S. Ito, H. Hashizume, N. Yanagi, H. Tamura, Advanced high-temperature superconducting magnet for fusion reactors: Segment fabrication and joint technique, Fusion Engineering and Design 136 (2018) 239-246. https://doi.org/10.1016/j.fusengdes.2018.01.072

[17] N. Yanagi, S. Ito, Y. Terazaki, K. Natsume, H. Tamura, S. Hamaguchi, T. Mito, H. Hashizume, J. Morikawa, Y. Ogawa, Feasibility of HTS magnet option for fusion reactors, Plasma and Fusion Research 9 (2014) 1405013-1405013. https://doi.org/10.1585/pfr.9.1405013

[18] W. Fietz, S. Fink, R. Heller, P. Komarek, V. Tanna, G. Zahn, G. Pasztor, R. Wesche, E. Salpietro, A. Vostner, High temperature superconductors for the ITER magnet system and beyond, Fusion engineering and design 75 (2005) 105-109. https://doi.org/10.1016/j.fusengdes.2005.06.198

[19] V. Braccini, A. Xu, J. Jaroszynski, Y. Xin, D. Larbalestier, Y. Chen, G. Carota, J. Dackow, I. Kesgin, Y. Yao, Properties of recent IBAD-MOCVD coated conductors relevant to their high field, low temperature magnet use, Superconductor Science and Technology 24 (2010) 035001. https://doi.org/10.1088/0953-2048/24/3/035001

[20] L. Bromberg, M. Tekula, L. El-Guebaly, R. Miller, A. Team, Options for the use of high temperature superconductor in tokamak fusion reactor designs, Fusion Engineering and Design 54 (2001) 167-180. https://doi.org/10.1016/S0920-3796(00)00432-4

[21] H. Hashizume, S. Kitajima, S. Ito, K. Yagi, Y. Usui, Y. Hida, A. Sagara, Advanced fusion reactor design using remountable HTc SC magnet, J. Plasma Fusion Res. Ser. 5 (2002) 532-536.

[22] N. Yanagi, S. Ito, Y. Terazaki, Y. Seino, S. Hamaguchi, H. Tamura, J. Miyazawa, T. Mito, H. Hashizume, A. Sagara, Design and development of high-temperature superconducting magnet system with joint-winding for the helical fusion reactor, Nuclear Fusion 55 (2015) 053021. https://doi.org/10.1088/0029-5515/55/5/053021

[23] N. Yanagi, T. Mito, R. Champailler, G. Bansal, H. Tamura, A. Sagara, Design progress on the high-temperature superconducting coil option for the heliotron-type fusion energy reactor FFHR, Fusion Science and Technology 60 (2011) 648-652. https://doi.org/10.13182/FST60-648

[24] A. Sagara, J. Miyazawa, H. Tamura, T. Tanaka, T. Goto, N. Yanagi, R. Sakamoto, S. Masuzaki, H. Ohtani, Two conceptual designs of helical fusion reactor FFHR-d1A based on ITER technologies and challenging ideas, Nuclear Fusion 57 (2017) 086046. https://doi.org/10.1088/1741-4326/aa6b12

[25] A. Sagara, H. Tamura, T. Tanaka, N. Yanagi, J. Miyazawa, T. Goto, R. Sakamoto, J. Yagi, T. Watanabe, S. Takayama, Helical reactor design FFHR-d1 and c1 for steady-state DEMO, Fusion Engineering and Design 89 (2014) 2114-2120. https://doi.org/10.1016/j.fusengdes.2014.02.076

[26] F.K. Ariyo, M. Omoigui, Investigation of the damping of electromechanical oscillations using power system stabilizers (pss) in nigerian 330 kv electrical network, Electrical and Electronic Engineering 2 (2012) 236-244. https://doi.org/10.5923/j.eee.20120204.10

[27] S. Helmy, A.S. El-Wakeel, M.A. Rahman, M.A. Badr, Mitigating Subsynchronous resonance torques using dynamic braking resistor, Journal of Energy and Power Engineering 6 (2012) 833-839.

[28] A. Rahim, A. Al-Sammak, Optimal switching of dynamic braking resistor, reactor or capacitor for transient stability of power systems, IEE Proceedings C (Generation, Transmission and Distribution), IET, 1991, pp. 89-93. https://doi.org/10.1049/ip-c.1991.0011

[29] J. Usman, M.W. Mustafa, G. Aliyu, Design of AVR and PSS for power system stability based on iteration particle swarm optimization, International Journal of Engineering and Innovative Technology (IJEIT) 2 (2012).

[30] D. Connolly, A Review of Energy Storage Technologies: for the integration of fluctuating renewable energy, (2010).

[31] R. Patel, T. Bhatti, D. Kothari, Improvement of power system transient stability by coordinated operation of fast valving and braking resistor, IEE Proceedings-Generation, Transmission and Distribution 150 (2003) 311-316. https://doi.org/10.1049/ip-gtd:20030301

[32] S. Zhang, F.L. Luo, Power system stabilizer based on improved simple adaptive control, 2008 3rd IEEE Conference on Industrial Electronics and Applications, IEEE, 2008, pp. 908-913.

[33] P.F. Ribeiro, B.K. Johnson, M.L. Crow, A. Arsoy, Y. Liu, Energy storage systems for advanced power applications, Proceedings of the IEEE 89 (2001) 1744-1756. https://doi.org/10.1109/5.975900

[34] X.Y. Chen, J.X. Jin, Y. Xin, B. Shu, C.L. Tang, Y.P. Zhu, R.M. Sun, Integrated SMES technology for modern power system and future smart grid, IEEE Transactions on Applied Superconductivity 24 (2014) 1-5. https://doi.org/10.1109/TASC.2014.2346502

[35] M. Sander, F. Brighenti, R. Gehring, T. Jordan, M. Klaeser, D. Kraft, R. Mueller, H. Neumann, T. Schneider, G. Stern, LIQHYSMES-Liquid H2 and SMES for renewable energy applications, International journal of hydrogen energy 39 (2014) 12007-12017. https://doi.org/10.1016/j.ijhydene.2014.06.008

[36] P. Zhao, J. Wang, Y. Dai, Capacity allocation of a hybrid energy storage system for power system peak shaving at high wind power penetration level, Renewable Energy 75 (2015) 541-549. https://doi.org/10.1016/j.renene.2014.10.040

[37] P. Mukherjee, V. Rao, Design and development of high temperature superconducting magnetic energy storage for power applications-A review, Physica C: Superconductivity and its applications 563 (2019) 67-73. https://doi.org/10.1016/j.physc.2019.05.001

[38] S. Ran, Gravity probe B: Exploring Einstein's universe with gyroscopes, NASA (2004) 26.

[39] N.J. Kelley, C. Wakefield, M. Nassi, P. Corsaro, S. Spreafico, D.W.V. Dollen, J. Jipping, Field demonstration of a 24-kV warm dielectric HTS cable, IEEE Transactions on Applied Superconductivity 11 (2001) 2461-2466. https://doi.org/10.1109/77.920361

[40] A. Sergeev, I. Golev, High-Temperature Superconducting Materials Based on Bismuth with a Low Critical Current, Materials Today: Proceedings 11 (2019) 489-493. https://doi.org/10.1016/j.matpr.2019.01.019

[41] A. Cansiz, E. Akyerden, The use of high temperature superconductor bulk in a co-axial magnetic gear, Cryogenics 98 (2019) 80-86. https://doi.org/10.1016/j.cryogenics.2019.01.008

[42] E. Gouda, S. Mezani, L. Baghli, A. Rezzoug, Comparative Study Between Mechanical and Magnetic Planetary Gears, IEEE Transactions on Magnetics 47 (2011) 439-450. https://doi.org/10.1109/TMAG.2010.2090890

[43] J. Rens, K. Atallah, S.D. Calverley, D. Howe, A Novel Magnetic Harmonic Gear, IEEE Transactions on Industry Applications 46 (2010) 206-212. https://doi.org/10.1109/TIA.2009.2036507

[44] M.M. Taşkın, A. Cansız, Design and Optimization of Generator for Narrow Body Commercial Aircraft, 2019 11th International Conference on Electrical and Electronics Engineering (ELECO), 2019, pp. 1007-1011. https://doi.org/10.23919/ELECO47770.2019.8990621

[45] B. Dianati, H. Heydari, S.A. Afsari, Analytical Computation of Air-Gap Magnetic Field in a Viable Superconductive Magnetic Gear, IEEE Transactions on Applied Superconductivity 26 (2016) 1-12. https://doi.org/10.1109/TASC.2016.2544832

[46] G. Malé, S. Mezani, T. Lubin, J. Lévèque, A Fast Analytical Method to Compute the Radial Flux Density Distribution in the Airgap of a Superconducting Inductor, IEEE

Transactions on Applied Superconductivity 21 (2011) 1114-1118.
https://doi.org/10.1109/TASC.2010.2096172

[47] J.L. Perez-Diaz, E. Diez-Jimenez, I. Valiente-Blanco, C. Cristache, M.-A. Alvarez-
Valenzuela, J. Sanchez-Garcia-Casarrubios, C. Ferdeghini, F. Canepa, W. Hornig, G.
Carbone, J. Plechacek, A. Amorim, T. Frederico, P. Gordo, J. Abreu, V. Sanz, E.-M.
Ruiz-Navas, J.-A. Martinez-Rojas, Performance of Magnetic-Superconductor Non-
Contact Harmonic Drive for Cryogenic Space Applications, Machines 3 (2015) 138-
156. https://doi.org/10.3390/machines3030138

[48] X. Yin, Y. Fang, P. Pfister, A Novel Single-PM-Array Magnetic Gear With HTS
Bulks, IEEE Transactions on Applied Superconductivity 27 (2017) 1-5.
https://doi.org/10.1109/TASC.2017.2672676

Superconductors - Materials and Applications
Materials Research Foundations 132 (2022) 194-210

Materials Research Forum LLC
https://doi.org/10.21741/9781644902110-11

Chapter 11

Superconducting Metamaterials and their Applications

M. Rizwan[1*], A. Usman[2], M. Zainab[3], A. Ayub[2], B. Tehreem[4]

[1]School of Physical Sciences, University of the Punjab, Lahore, Pakistan

[2]Department of Physics, University of Punjab, Lahore, Pakistan

[3]Department of Physics, University of Gujrat, Gujrat, Pakistan

[4]Austin Medical Assistant Training, New York, United States

*rizwan.sps@pu.edu.pk

Abstract

Superconductors have been in the field for more than a century now and superconducting metamaterials are a class of materials that have extremely low losses. Superconducting materials exhibit supreme properties such as currents that can last for twenty-six years and quantum phenomenon that makes it very suitable for making metamaterials. Superconductor composites are materials in which a material with a substantial permeability and negative effective permittivity material is combined with a superconductor which reduce losses in metamaterials at resonance frequency. There are many angles to explore and investigate in superconducting metamaterials. Novel applications of superconducting metamaterials are also briefly deliberated

Keywords

SQUIDS, Metamaterials, Split Ring Resonators (SRRs), Superconductors, Josephson Junction

Contents

1.   Superconducting materials

Superconductivity is a physical phenomenon that occurs in some materials when electrical resistance vanishes and magnetic field is excluded at a certain value of temperature. A superconductor possesses a critical temperature, under this value of temperature, the resistivity plunges to zero, on the other hand, regular metallic conductor, whose resistance have a direct relationship with temperature and deceases as temperature is reduced , even to near absolute 0 [39]. An electric current can flow endlessly through a circuit of superconducting wire because of zero resistivity.

Our goal here is to provide a summary of the new field of superconducting metamaterials. Superconductors have a range of electromagnetic properties that conventional metals don't have and these features can be used to generate near-perfect and innovative metamaterial structures. Zero direct current (DC) resistance, Meissner effect, pure diamagnetism, and macroscopic quantum events are three unique properties of superconductivity. Kamerlingh Onnes established the zero DC resistance characteristic in 1911, and this property of superconductors has enabled many applications of superconductors from power generation to battery storage [29]. Superconductors possess another distinctive feature that is full diamagnetic response in the existence of a heavy magnetic field. Currents flows to keep the magnetic field out of its interior, in the superconductors. The Meissner effect is a phenomenon that distinguishes superconductors from ideal conductors. Finally, the quantum mechanical character of electron in associated the superconducting causes macroscopic quantum phenomena [29].

A macroscopic quantum wave equation can be used to explain superconducting particles in some cases. The wave function in the wave equation has amplitude that can be translated into the density distribution of superconducting particles and a period that is cohesive in macroscopic proportions. Quantum interference and tunneling effects are produced by this phase coherence wave function, which are highly exceptional and specific to the state of superconductors. Characteristics of macroscopic quantum theory are DC/AC valuable source effects at tunneling barriers and fluxoid quantification [3].

The imaginary part of the conductance, which is substantially bigger in scale than the true part at limiting temperature and is highly frequency related ($\sigma_2 \sim 1/\omega$), dominates the alternating current wave particle duality of a superconductor. A superconductor's reaction is thus largely diamagnetic and inducing. The impedance is caused by screening flux that flow in the bulk of the material to keep it in the Meissner state. As a result of this Meissner effect, the electric current flows within the range of a depth of penetration of the surface [5].

In the two-fluid model, the depth of penetration is inversely proportional to the supercritical fluid density, as given by $\lambda^2 = m/(\mu_0 n_s e^2)$, where e represents the electrical charge and m stands for mass. The penetration depth is material dependent and has the range from around ten second to hundred second of nanometer range at absolute zero. The supercritical fluid density reliant on temperature up to a certain temperature, where it becomes zero and before that it linearly decreases with temperature [21]. As T_c is reached from below, the penetrating depth diverges, representing depth absorption of electromagnetic waves and the deprivation of the superconducting state.

Superconductor's inductance is owed to the stored energy outside and within the superconductor's magnetic field, and also the kinetic energy kept in the super current flow. The kinetic impedance of a tinny, narrow wires with cross-sectional aspects can be written as: $L_{Kin} = \mu_0 \lambda^2/t$, where t is the layer thickness [35]. The supercritical fluid density can be reduced to 0 by applying a magnetic field, or increasing the temperature to T_c, or by obtaining the critical current and thus the dynamic inductance can be divided. When a superconducting wire is used in an electrical circuit at this level, the kinetic inductance of the link might change dramatically causing the circuit to collapse[9].

Establishing an electrical series connection of a sample in which current I flows with measuring voltage V is the simplest method for determining its electrical resistance, V=IR, which is Ohm's law's formula for the sample's resistance. The resistance is 0 if the voltage is removed.

Superconductors can also retain a current with no applied voltage, which is used in superconducting electromagnetic fields like those used in (magnetic resonance imaging) MRI equipment. Current in superconducting circuits can sustain for years without degrading in any way according to various experiments. Experiments show that humans have a present lifetime of at minutest 100,000 years. The duration of a sustained current can become equal to the duration of universe, reliant on the wire architecture and temperature in superconductors [24]. In exercise, currents infused in superconductivity coils have lasted in superconducting gravimeters for even more than 26 years.

In normal conductors, an electric current can be characterized as a liquid of electrons flowing over a dense ionic structure. Electrons regularly collide with the particles in the lattice, but lattice ion's oscillating kinetic energy is provided to current as a result of collisions which is turned into heat. Because of which, the power that the current conveys is constantly dispersed. This is the electrical resistance and Heat transfer phenomenon [24]. When the temperature T is dropped under a threshold temperature T_c, superconductivity features occur in superconducting elements. The temperature at which this temperature range is reached varies depending on the material. Critical temperatures for conventional superconductors typically range from roughly 20 K to less than 1 K. The temperature required for solid mercury, for example, is 4.2 K. The maximum critical temperature for a standard superconductor discovered as of 2015 is 203K for H_2S despite the need for high pressures of around 90 GPa [42]. Crystalline superconductors have much higher critical temperatures than other superconductors: $YBa_2Cu_3O_7$, one of the very first crystalline superconductors identified, has a critical temperature > 90 K, while cuperates with mercury have critical temperature of over 130 K. In superconductors the ordering of phase transition have been a point of difference for a long time. Researches have shown that the phase transition is in fact 2nd order thus representing that no heat capacity is present. However, because the superconductivity state has lowered entropy under the critical temperature than the normal phase there is latent heat in the incident electromagnetic field. When the magnetization is raised outside the specific field the following phase transition causes the rate of the superconductivity material to drop [22].

2.   Metamaterials

Metamaterials are often made up of "atoms" with electromagnetic responses that have been designed. Artificial atom characteristics are frequently created to create non-trivial effective permeability and permittivity values for a matrix of two atoms. Comparative permittivity and permeabilities which is less than 1, near to zero or negatives are examples of such quantities. We will consider the generalization capabilities of metamaterials produced of traditional "atomic" systems, which are often used in the initial metamaterials literature. Wires are used in traditional metamaterials to manipulate the actual plasma frequency of the substance to modify its dielectric characteristics[10]. Split-ring resonators (SRRs) employ their magnetization to establish frequency range with sub-unity, minus, or near-0 magnetic properties. One of the major drawbacks of traditional metamaterials is considerable losses.

Ohmic losses pose a limitation in the RF-THz frequency range performance of metamaterials. Unlike ordinary metals, Split ring resonators and superconducting wires can be significantly reduced in size while keeping their low loss features. Losses grow as ρ/r^2 and ρ/tl, correspondingly, as the length of regular electrical connections and SRRs is reduced, where r represents cable radius , the dimension of the SRR is represented by l,

Superconductors - Materials and Applications   Materials Research Forum LLC
Materials Research Foundations 132 (2022) 194-210   https://doi.org/10.21741/9781644902110-11

and thicknesses of the material is t, where is ρ is the resistance of the metal, that makes up the SRR [12].

Increasing the plasma frequencies of an artificial insulator by decreasing the typical wire coil diameter will result in a large jump in losses. Losses increase as the typical metal SRR proportions decrease, and the frequency bandwidth of μ_{eff}(f) is less than 0 and finally disappears. Since the resistivity of superconductivity wires and SRRs is usually minor and the electromagnetic responses are controlled by the responsive impedance, these negative consequences do not occur. When the generated currents reach the critical conductance of the value $(J_c \sim 10^6 - 10^9 A/cm^2)$ or proportions match the coherence length, or superconductors fail.

2.1   Low loss metamaterials

Metamaterials with such a negative refractive index (NIR) must be manufactured very minute in comparison to the wavelength and with very low distortion for new applications. Spectral efficiency of ultra-small dipole antennas can be enhanced using metamaterials with physical dimensions in between 1 mm operating at 10 GHz as proposed by Ziolkowski and Kipple. Some metamaterials require elements on the order of tens of micrometers (wires and split-ring resonant frequencies). Significant losses will thrash the intended NIR behavior, thus it's important to check if current metamaterials concepts can be pared back to the needed dimensions [7]. Scaling of ordinary normal metal metamaterial constructed from split-ring resonators (for reverse permeability) and wires (for negative permittivity) can be performed in this way.

2.2   Scaling of SRR properties

We will now deliberate what consequences happen as we try to compress normal metals to attain the above-mentioned metamaterial proportions. Currently, relatively chunky plated metals which have thickness between 25 and 50 m that are placed on uncompressed dielectric materials are used to produce splitting resonators (SRRs) which operates at frequency 10 GHz (such as FR4 and G10) [38]. The outside dimensions of these SRRs are about 2.5 mm, and the gaps widths are of g = 300–500 m.

If SRR is scaled down to a degree of ten or more, than the efficient relative magnetic permeability of an SRR arrays can be calculated as follows, the SRR's resonant frequency is; $\omega_0 = c_0 \sqrt{\frac{3l}{\pi r^3 ln\frac{2c}{d}}}$ the area of $Re[\mu_{eff}]$ <0 is immediately above all this resonant frequency. Where l is the cubic SRR element's lattice gap, c_0 is the velocity of light and inner radius of the inner SRR is given by r. The SRR's resonant frequency will scale nearly as $\omega_0 \sim 1/r$ as radius is decreased because the lattice parameter is increased with the radius l~r, keeping ratio c/d remains constant. This equation is missing the scaling dependency given by separation g that explains capacitive effect. Capacitive effect of the separation g can be made very small in order to improve capacitance, while keeping the split ring resonator resonant frequency constant at 10 GHz, as SRR lessens [37].

Superconductors - Materials and Applications   Materials Research Forum LLC
Materials Research Foundations 132 (2022) 194-210   https://doi.org/10.21741/9781644902110-11

2.3 Scaling of wire array properties

For frequencies which are smaller than the plasma frequency, the real component of the wires array's effective permittivity is low. The plasma border will increase nearly as ω_p ~1/a as the cable arrays lattice parameter decreases. At 10 GHz, this scalability will retain the intended μ_{eff} <0 limit.

$$\mu_{eff} = 1 - \frac{\omega_p^2}{\omega(\omega + \frac{i\varepsilon_0 a^2 \omega_p^2}{\pi r^2 \sigma})} \qquad \text{Eq.1}$$

This term gives the complex effective relative electric permittivity; the term $\frac{\mu_0 a^2 \omega_p^2 \rho}{\pi r^2}$ gives the loss term, where $\sigma = 1/\rho$ is the typical wire coil conductivity. Because ω_p ~1/a, this results in an electrical resistivity that scales as ρ/r^2 as the wire radius decreases. Downsizing of a material causes the losses to increases as they have an inverse square dependence with the radius [17]. As the proportions of traditional normal metal SRR and wire arrangements are gradually scaled down to meet the sizes necessary for non-trivial applications their characteristics suffer a severe rise in losses. This makes regular metals unsuitable for any application requiring a metamaterial's dimension to be reduced by an order of magnitude or more while starting with least tolerable losses. Superconducting metamaterials are one obvious option for resolving this challenge [23].

3. Novel superconducting metamaterial implementations

There are three hallmark that characterize superconductors properties, such as zero DC resistance, Meissner effect macroscopic quantum effects and diamagnetism. The zero DC resistance was exposed by Kamerlingh Onnes in 1911 and has many significant advantages [26]. The 2nd trademark is that superconductors fundamentally complete diamagnetic response under the existence of a static magnetic field. When the material becomes superconductor in superconducting state, current is developed from its interior to eliminate the magnetic field. This phenomenon is called Meissner effect [30]. This phenomenon distinguished the superconductors from the perfect conductors.

'LowT_c' superconductors are also referred as traditional superconductor. Their transition temperature lies in the range of mK to 25K. Transition temperature of 'highT_c' superconductors are up to 150K. An intermediate temperature ranges from 5K to 50K range, number of interesting superconductors have been discovered. Superconducting magnets are created by technological superconductor include Nb and different alloys. This superconducting magnets are used for magnet resonance imaging and particle accelerators [4].

Metamaterials are man-made materials that are customized possess characteristics that are not present in natural materials. For example, negative refraction. In nature every known element has positive refraction index. Superconductors are potential candidate for application in metamaterials or for the photonic crystal because of their intrinsic properties [30]. Superconducting metamaterial is made from the superconducting metals. Ultra-low losses, their compact structure, their degree of non-linearity and tunability, magnetic flux quantization, quantum effects, strong diamagnetism and Josephson Effect are all the properties of metamaterials brought by superconductivity [36].

There are many angles of research to explore in superconducting metamaterials. Their unique properties make them an excellent candidate for understanding of the landmark prediction of metamaterials theory including hyper lensing, transformation optics, near perfect lens, evanescent wav amplification [5].

Along with (SRRs) superconducting splits rings and wires, a number of unique superconducting metamaterial applications have been developed.

3.1   Ferromagnet- superconductor composites

Ferromagnetic resonance aids us to have negative real relative permeability of gyromagnetic materials for frequencies greater than resonance frequency. The $Im[\mu_{eff}]$ of the permeability is enormous at the resonant frequency, limiting the applications of such $Re[\mu_{eff}] < 0$ materials. Combining ferromagnetic substance with a superconductor can limit losses while simultaneously introducing a conducting circuit with $Re[\mu_{eff}] < 0$.

A negative-index region in the 90 GHz range was generated by constructing a super lattice sheet made up of high-temperature superconductors and nanostructured ferromagnetic layers [13]. Despite the fact that the (Im[n]) was of equivalent scale to Re[n]<near a 3 T applied electric field at 90 Gigahertz, the material showed Re[n]<0.

3.2   DC magnetic superconducting metamaterials

Wood and Pendry presented and investigated the idea of a DC magnetic shield. The main idea is to cause disruption in a solid diamagnetic superconducting material into smaller components and arrange them in such a way that the magnetic reaction can be adjusted. The magnetic field distributions remain unaffected, and the cloak protect a certain area from external Dc field and uses superconducting plates to give diamagnetic reaction, thus allowing the radial aspect of effective permeability to range from 0 to 1 [32]. Paramagnetic compounds were suggested to increase the tangential elements of magnetic permeability so. A cloak was produced using an assembly of lead thin film sheets, with an experimental illustration of the first step ($\mu_r < 1$). The DC magnetic cloak concept has now been revised theoretically, with the suggestion that it can be realized with large enough temperature superconducting thin sheets [16].

3.3   SQUID metamaterials

The split-ring resonator has a suitable quantum analogue: a SQUID (superconducting quantum interference devices). In reality, the Radio frequency SQUID, which was invented in the 1960s, is basically a quantum SRR with a Josephson junction replacing the classical capacitors. Its original function was to use as a fluid flow to frequency transducers and

Superconductors - Materials and Applications   Materials Research Forum LLC
Materials Research Foundations 132 (2022) 194-210   https://doi.org/10.21741/9781644902110-11

measure tiny RF magnetic fields [20]. They base their calculations on the assumption that the energy levels of SQUIDS are quantized and that the photon in radio wave range only interacts with SQUID current in the lowest lying regions. The material will have a negative effective permeability because of the small detuning photon frequency in the microwave region that is the transitions from the initial state to the 1st excited singlet state. On the other hand, non-zero de-phasing rate reduces the frequency area of negative permeability, and negative permeability disappears for de-phasing rates greater than a significant level. Lazarides and Tsironis proposed a theory of RF SQUIDs operating with classical electromagnetic radiation. The Josephson connection was regarded as a parallel connection of Josephson impedance, resistance and capacitance in two-dimensional arrays of RF SQUIDs. A simple RF SQUID can have a high diamagnetic reaction near resonance [11]. The permeability is periodic as a response to applied magnetics flux, and it is repressed by applied fields that stimulate current flow in the SQUID that are greater than the Josephson junction's critical current. Rakhmanov et al. considered that 1-D network of superconducting islands can operate as qubits. The network of superconducting islands can form a quantum optical crystal that can sustain a range of wave propagation excitations when it interacts with traditional electromagnetic waves. To conduct parametric enhancement of signal transmission a comparable technique reliant on a SQUID power line was created [40].

4. Superconducting photonic crystal

Photonic crystal were first discussed by Yablonovitch and John. Photonic crystals have been constructed from the modulated dielectric function. Photonic crystal fall outside the domain of metamaterial but we shall consider them in it [6]. Most of the work of superconducting photonic crystal is on the metamaterials. Initial work on superconducting photonic crystals introduced the concept about propagation modes in anisotropic cuprite superconducting photonic crystals and a novel temperature dependent gap similar to phonon-polariton gap which is also called a new spectroscopic feature in superconducting Photonic crystals [15].

Photonic crystals can control the flow of electromagnetic waves and thus became the objective of various theoretical and experimental studies. In the past, due to the advantages of photonic crystals, superconductors have been incorporated in them. The main advantage of photonic crystal is that it can replace metals because it has the ability to reduced damped electromagnetic waves. Superconducting photonic crystal offers large reflection bands that can be scaled down by the temperature. The response of superconducting photonics crystals is reliant on London penetration depth which is temperature and magnetic field dependent [25]. We can compute the band structure of photonic crystal with superconducting elements via the process of plane wave expansion. It is an artificial media with periodic dielectric function. Photonic crystals have the various modern applications.

At small frequencies, only metal contained in photonic crystals maintains the dielectric contrast. But in this case at frequency ω, photonic gaps are obtained for high dielectric

Superconductors - Materials and Applications   Materials Research Forum LLC
Materials Research Foundations 132 (2022) 194-210   https://doi.org/10.21741/9781644902110-11

contrast. Photonic crystal comprising of superconducting elements are used for frequencies smaller than superconducting gap.

We consider London approximation for superconducting particles:

$$\delta_L = (\frac{mc^2}{4\pi n e^2})^{\frac{1}{2}} \gg \xi \qquad \text{Eq.2}$$

ξ is coherence length. It is a simple relation for current density.

$$At \frac{T_c - T}{T_c} \ll 1 \text{ and } \hbar\omega \ll \Delta \ll T_c$$

$$J(r) = [-\frac{c}{4\pi\delta_L^2} + \frac{i\omega\sigma}{c}] A(r) \qquad \text{Eq.3}$$

Substitute equation (3) in wave equation

$$-\nabla^2 E = \frac{1}{c^2}\epsilon_0 \frac{\partial^2 E}{\partial t^2} - \frac{4\pi}{c^2}\frac{\partial J(r)}{\partial t} \qquad \text{Eq.4}$$

As we know that E $(r,t) = E_0(r)e^{i\omega t}$, E $= \frac{i\omega}{cA}$, E (r,t) is harmonic time variation of electric field.

Calculations are for two component photonic crystals

$$-\nabla^2 E = \frac{\omega^2}{c^2}[\epsilon_o + (\epsilon_s(\omega) - \epsilon_o)\sum\eta \ (r \in S)]E \qquad \text{Eq.5}$$

$\eta \ (r \in S) = 1$ If r is inside the superconductor element S otherwise it is equals to zero.

$\epsilon_s(\omega)$ is the dielectric function of superconductor and ϵ_o is dielectric component of surrounding medium. We use the approximation $\epsilon_s(\omega) = 1 + \frac{c^2}{\delta_L^2\omega^2}$. Summation in equation 5 characterize the position of superconducting particles over all the lattice nodes [8]. So we are going to compute the band structure of photonic crystals comprising of an unlimited array of superconducting cylinder. So we reduce the Maxwell equation to Eigen value problem.

$$\sum_{G'}(\delta_{GG'} (K + G)^2 - \frac{P^2}{c^2}M_{GG'})E_k(G') = \frac{\omega^2}{c^2}E_k(G), \qquad \text{Eq.6}$$

Here P $= \frac{c}{\delta_L}$; G $= i_1 G_1 + i_2 G_2$ is the reciprocal lattice vector foe 2D square lattice. Here i_1 and i_2 are the arbitrary integer and G_1 and G_2 are 2D orthogonal vectors. $E_k(G)$ is the

Fourier components of electric field. The matrix $M_{GG'}$ is related to Fourier transform of dielectric function of superconducting photonic crystal:

$$(G - G') = \delta_{GG'} + \frac{p^2}{\omega^2} M_{GG'} \qquad \text{Eq.7}$$

Where

$$M_{GG'} = f\epsilon(f - 1), G = G' \qquad \text{Eq.8}$$

$$M_{GG'} = 2f(\epsilon - 1)\frac{J_1(|G-G'|r)}{|G-G'|r}, G \neq G' \qquad \text{Eq.9}$$

Here 'f' is the filling factor of superconductor. $f = \frac{S_{supercondu}}{S} = \frac{\pi r^2}{a^2}$. S is the total area occupied by lattice. A is spacing of lattice. 'r' is cylindrical radius. And J_1 is the Bessel function. The periodicity of dielectric function can be accomplished by introducing an array of essential elements with dielectric ϵ_1 in the background medium which is characterized by ϵ_2. A photonic gap exist only in the range of 0.23< f < 0.44. Where 'f' is the filling factor. Widest gap ($\Delta = 0.35$) exists only when f = 0.32. Superconducting materials have a negligible part of dielectric function in comparison to normal metals which causes dissipation of electromagnetic waves, when these materials are combined with photonic crystal with superconducting particle [2].

5.   Thin film superconducting metamaterial

Superconducting thin films in comparison to normal metals can produce the lossless large inductance values. Resonant structures of metamaterials can be built with large inductance value at low frequencies because of the ability of superconducting thin films providing less losses and large inductance values. 3-D structures have been utilized in past to produce the non-natural magnetism below the 100 MHz [41]. By employing thick normal metal wire, 2-D spiral resonators are used to produce the negative effective permeability atom in sub-gigahertz domain. So, superconducting thin film spiral enhanced the inductance and also have the low losses. Superconducting spiral metamaterials have been developed with Niobium thin film. Strong tuning ability is observed as transition temperature is reached. Application of such metamaterials are compact RF resonator application, MRI (magnetic resonance imaging) and near field imaging [18].

6.   Advantages of metamaterials

Atoms being engineered by electromagnetism are known as metamaterials. The purpose behind setting up characteristics of artificial atoms is to optimize the permittivity and permeability of lattice atoms. The opt6mized values range from being negative, less than 1 or close to zero [30, 31].

Superconductors - Materials and Applications   Materials Research Forum LLC
Materials Research Foundations 132 (2022) 194-210   https://doi.org/10.21741/9781644902110-11

6.1   Compact superconducting materials

At microwave frequencies ranging from 100's of MHz to 100's of GHz , superconductors hold very small values of surface resistance. Normally at room temperature superconductors have surface resistance values for Niobium observed at 2K a 1 GHz and are given in nano-Ohm range and at high temperature the scale range is 100µ at 77k and 10 GHz. The superconductors can handle these scaling limitations by permitting these small losses explained earlier [14]. The kinetic inductance can be increased by decreasing the cross-sectional area of superconductors in a such way that single or the two sides are on the scale of magnetic penetration depth λ, as talked recently, the dissipation-less supercurrent flow of inductance is related to kinetic energy.

The Josephson Effect becomes active if we reduce the cross sectional area to an extent that the width lies under the scale of coherence length ξ. Another way to achieve the subject matter is by generating a tapered tunnel barrier in the superconductor .Therefore, inductance could be raised due to currents flowing through Josephson junction. Such configurations of inductance give rise to the superconducting metamaterials which are even more miniature in comparison to the standard metal based metamaterials.

6.2   Tuneability and nonlinearity

Inductance of superconductors is greatly tunable along with being nonlinear because of their eccentric quantum mechanical properties. The superfluid density could be seen as an effective function of current and applied magnetic fields (both Dc and RF). This function leads to the modification in Josephson and kinetic inductance. The following research indicates the utilization of temperature dependent tuning of negative permeability space of SRRs. DC current can also be exploited to control the kinetic inductance of superconductors [32].

Superconductors also bear the potential of quantized magnetic vortex excitations, in turn when these are agitated by RF currents, producing inductance and loss. If the currents flowing in SRRs are changed, we can tune the negative permeability region in the superconducting materials. This happens because of the RF magnetic vortices that enter SRRs honed inside corners of patterned structure. The confirmation was realized by images of radio frequency current flow at a high temperature superconducting SRR. It has also been observed that by adding DC magnetic vortices to the superconducting SRRs, the negative permeability region indicates frequency shifts [9]. It lead to the hypothesis that the reason behind this frequency shift is a nonuniform delivery of magnetic flux that enters and leaves split ring resonators as the field is stormed. Microscope images of flux penetration through magneto-optical imaging added to the evidence of highly inhomogeneous nature of flux penetration.

An overture way to produce the photonic crystals is by magnetic vortex lattice in superconductors, between the non-superconducting cores and superconducting bulk [24]. The brief description of these materials is described below. for currents and applied fields, the Josephson inductance is considered a stronger function.In a transmission line geometry, the tunning of Josephson inductance was coming in consideration theoretically by Salehi. The tunability by external fields was examine by Duetal of SQUID inductance. The Josephson inductance generating in an extended Josephson junction is considered very sensitive to the vertox [28].

6.3   Implementations of superconducting metamaterials

Although the superconductors are good conductors, superconductors cannot maintain a cryogenic environment, and in this developing age it's a need to maintain this environment. Cryogenic is like 'temperature below the freezing point of water'. With the help of this we can transfer and receive the signals at room temperature. If we say we can use liquid cryogens then it will not be appropriate as the liquid cryogens such as nitrogen and helium are expensive and are troubled. The invention of high-Tc superconductors made the closed cycle and cryocooler system unbelievably small and good in work and cheap in price. We can operate these systems for at least five years with warranty and adjust the heat load due to microwave input and transmission lines output to room temperature [16].

The factors that can affect the behavior of metamaterials and superconducting properties are temperature variations, stray magnetic field or strong RF power. Superconductors are prone to suffer by different ecological effects. For a well-grounded result, one needs to carefully control the temperature and high quality magnetic shielding of superconducting devices [43].

7.   Novel applications

Superconducting metamaterials have a very exciting future especially in the low loss category of materials. These materials are superior to normal metals in guiding electromagnetic waves over long distances [16]. Superconductivity low losses and combination of Josephson effects leads to tunable devices with large bandwidth that are very pivotal for quantum interreference devices realization. RF-SQUID arrays with bistability and muti stability are also very exciting [33]. Compact superconducting metamaterials can be further scaled down die to introduction of josephson and SQUID in SRRS. Superconducting metamaterials have also been considered for quantum optics. Superconducting quantum metamaterials have extensive application as microwave photons detector, quantum birefringence and in detection super radiant phase transition. These quantum bits have strong effective coupling to external field and thus opens opportunities for making artificial quantum structures [19].

Superconducting metamaterials have extensive applications such in SQUIDS. There is a room for designing various kinds of resonant structures by modification of left-handed and right-handed media propagation [1]. Research by Engheta showed that resonant structures created by covering two materials in inverse senses of modified group of resonant structures. Before a wave achieves another reflection boundary, it propagates in a direction normal to plates undergoing a mixed forward and reverse phase winding [27]. According to this situation, a wave can also produce a resonance condition by undergoing a net phase shift of 0 radians. Negative or positive multiple of 2π radians net phase shift is possible which creates resonance. An ultra-compact resonator is created with a whole dimension no more restricted by wavelength of the resonant wave.

Wang and Lancaster were the first among who established the understanding of such a kind of ultra-compact resonator of superconductors in which a dual transmission line structure was executed in a coplanar waveguide geometry. These constructions successfully led to the synthesis of applicable backward-wave microwave devices. Salehi *et al* also researched their superconducting versions. The superconducting subwavelength resonator had been used with cuperates superconducting films. The superconductors have been found exclusively beneficial for application in ultra-compact resonates as in other cases Q can be reduced by ohmic losses of a profound sub-wavelength structure. It will lead to inefficiency of the device's performance. The quality factors related to negative order resonances had been of order 3000 at 30 K. Whereas positive order resonance was below 400 [31].

Incorporating Josephson inductance in a dual transmission line which is superconducting can also lead to an increase in nonlinearity. It creates a dispersion relation that can be tuned thus generating waves in the structure. Taking into account a metamaterial consisting of a Josephson qubit array results in a various attractive prognosis. It comprises quantum photonic crystal development which borrows its characteristics from quantum states available to qubits. A 1-D SQUID array nonlinear transmission line is utilized for a parametric amplifier which can be tunes for microwave signals in 4-8GHz range. It then releases upto 28 dB of gain. Such an amplifier finds its utilization in signal detection in microwave qubits working at low temperature. it also can squeeze quantum noise [34].

The research in the field of metamaterials is increasing faster especially in reference to sensor applications. For example, high quality resonance aided with asymmetrically split ring resonators. Also, it can uncover analytes with low concentration e.g. sugar or hydrogen using nanoscales antennas. By focusing on the example of a single molecular layer of graphene, transmission of metamaterials could be enhanced many times. Moreover, solutions based on harnessing light energy would be possible by means of plasmonic metamaterial nanostructures. But it requires a significant limitation in physical thickness and increased efficiency in absorber layers of photovoltaic cells [4].

Conclusion

Superconductors are material that exhibit zero DC resistance, Meissner effect and have inductance and expel magnetic field. Materials with negative effective permeability and permittivity are called metamaterials, and such materials require to be less than the wavelength the scaling down of metamaterials can be achieved through reducing lattice parameter and radius of splint ring resonator, and tuning the resonance frequency, but this produce very considerable losses. These losses are controlled by introducing superconducting metamaterials and Josephson effect in metamaterials. Superconducting composites, DC magnetic superconducting materials, photonic crystals and thin film based superconducting metamaterials which are achieved by incorporating superconducting meta

Superconductors - Materials and Applications
Materials Research Foundations 132 (2022) 194-210

Materials Research Forum LLC
https://doi.org/10.21741/9781644902110-11

materials in metamaterials and these materials have a very bright future and applications in MRI as quantum bits in detection of quantum birefringence and many more. Superconductor composites are materials in which a material with a substantial permeability and negative effective permittivity is combined with a superconductor which reduce losses in metamaterials at resonance frequency. Superconducting quantum interreference devices (SQUIDS) that are quantum analogue of SRRs and they exhibit negative effective permeability for a first transition.

References

[1] A. Abdumalikov Jr, O. Astafiev, A.M. Zagoskin, Y.A. Pashkin, Y. Nakamura, J.S. Tsai, Electromagnetically induced transparency on a single artificial atom, Phys.Rev. Lett. 104 (2010) 193656- 193601. https://doi.org/10.1103/PhysRevLett.104.193601

[2] A.H. Aly, S.-W. Ryu, H.-T. Hsu, C.J. Wu, THz transmittance in one-dimensional superconducting nanomaterial-dielectric superlattice, Mater. Chem. Phys. 113 (2009) 382-384. https://doi.org/10.1016/j.matchemphys.2008.07.123

[3] S. Anlage, H. Snortland, M. Beasley, A current controlled variable delay superconducting transmission line, IEEE Trans. Magn.25 (1989) 1388-1391. https://doi.org/10.1109/20.92554

[4] S.M. Anlage, The physics and applications of superconducting metamaterials, J. Optic. 13 (2010) 023995 -024001. https://doi.org/10.1088/2040-8978/13/2/024001

[5] F. Auracher, T. Van Duzer, RF impedance of superconducting weak links, J. App.Phys. 44 (1973) 848-851. https://doi.org/10.1063/1.1662270

[6] C. Buzea, T. Yamashita, Review of the superconducting properties of MgB2, Supercond. Sci. Technol. 14 (2001) R110- R115. https://doi.org/10.1088/0953-2048/14/11/201

[7] F. Capolino, Theory and phenomena of metamaterials, CRC press, 2017 https://doi.org/10.1201/9781420054262

[8] Y. -B. Chen, C. Zhang, Y.-Y. Zhu, S. -N. Zhu, N.-B. Ming, Tunable photonic crystals with superconductor constituents, Mater. Lett. 55 (2002) 12-16. https://doi.org/10.1016/S0167-577X(01)00610-3

[9] S. Chui, L. Hu, Theoretical investigation on the possibility of preparing left-handed materials in metallic magnetic granular composites, Phys. Rev.iew B. 65 (2002) 14400- 144407. https://doi.org/10.1103/PhysRevB.65.144407

[10] T.J. Cui, D.R. Smith, R. Liu, Metamaterials, Springer., Boston, 2010, pp. 370-375

[11] C. Du, H. Chen, S. Li, Stable and bistable SQUID metamaterials, J. Phys. Cond. Matter. 20 (2008) 345218- 345220. https://doi.org/10.1088/0953-8984/20/34/345220

[12] N. Engheta, R.W. Ziolkowski, Metamaterials: physics and engineering explorations, JWS., 2006, pp. 430-440 https://doi.org/10.1002/0471784192

Superconductors - Materials and Applications   Materials Research Forum LLC
Materials Research Foundations 132 (2022) 194-210   https://doi.org/10.21741/9781644902110-11

[13] W. Gillijns, A.Y. Aladyshkin, A.Silhanek, V. Moshchalkov, Magnetic confinement of the superconducting condensate in superconductor-ferromagnet hybrid composites, Phys. Rev. B. 76 (2007) 060500- 060503. https://doi.org/10.1103/PhysRevB.76.060503

[14] H. Iwasaki, Y. Nakayama, K. Ozutsumi, Y. Yamamoto, Y.Tokunaga, H. Saisho, T. Matsubara, S. Ikeda, Compact superconducting ring at Ritsumeikan University, J. Synchrotron Radiat. 5 (1998) 1162-1165. https://doi.org/10.1107/S090904959701830X

[15] J.D. Joannopoulos, S.G. Johnson, J.N. Winn, R.D. Meade, Molding the flow of light, second ed., Princeton Univ. Press., Princeton, NJ [ua], 2008, pp. 200-304

[16] P. Jung, A.V. Ustinov, S.M. Anlage, Progress in superconducting metamaterials, Supercond. Sci. Technol. 27 (2014) 072995- 073001. https://doi.org/10.1088/0953-2048/27/7/073001

[17] V. Kantsyrev, L. Rudakov, A. Safronova, A. Velikovich, V. Ivanov, C. Coverdale, B. Jones, P. LePell, Ampleford, D.,Deeney, C., Properties of a planar wire arrays Z-pinch source and comparisons with cylindrical arrays, High energy density Phys. 3 (2007) 136-142. https://doi.org/10.1016/j.hedp.2007.02.009

[18] C. Kurter, S.M. Anlage, Superconductivity takes the stage in the field of metamaterials, SPIE Newsroom. 10 (2010) 002540-002543. https://doi.org/10.1117/2.1201002.002543

[19] N. Lambert, Y.-N. Chen, R. Johansson, F. Nori, Quantum chaos and critical behavior on a chip, Phys.Rev.B. 80 (2009) 165300-165308. https://doi.org/10.1103/PhysRevB.80.165308

[20] N. Lazarides, G. Neofotistos, G.Tsironis, Chimeras in SQUID metamaterials, Phys. Rev. B. 91 (2015) 054298- 054303. https://doi.org/10.1103/PhysRevB.91.054303

[21] S. -C. Lee, C.-Y. Lee, S.M. Anlage, Microwave nonlinearities of an isolated long YBa2Cu3O7− δ bicrystal grain boundary, Phys. Rev. B. 72 (2005) 024520- 024527. https://doi.org/10.1103/PhysRevB.72.024520

[22] F. Magnus, B. Wood, J. Moore, K. Morrison, G. Perkins, J. Fyson, M. Wiltshire, D. Caplin, L. Cohen, J. Pendry, A dc magnetic metamaterial, Nature ater. 7 (2008) 295-297. https://doi.org/10.1038/nmat2126

[23] O.T. Naman, M.R. New Tolley, R. Lwin, A. Tuniz, A.H. Al Janabi, I. Karatchevtseva, S.C. Fleming, B.T. Kuhlmey, A. Argyros, Indefinite Media Based on Wire Array Metamaterials for the THz and Mid IR, Adv. Optic. Mater. 1 (2013) 971-977. https://doi.org/10.1002/adom.201300402

[24] T. Nurgaliev, Modeling of the microwave characteristics of layered superconductor/ferromagnetic structures, Physica C: Supercond. 468 (2008) 912-919. https://doi.org/10.1016/j.physc.2008.04.001

[25] C.R. Ooi, T.A. Yeung, C. Kam, T. Lim, Photonic band gap in a superconductor-dielectric superlattice, Phys. Rev. B. 61 (2000) 5900-5920. https://doi.org/10.1103/PhysRevB.61.5920

[26] T.P. Orlando,K.A. Delin, C.J. Lobb, Foundations of applied superconductivity, Phys. Today. 44(1991) 100-109. https://doi.org/10.1063/1.2810145

[27] J.B. Pendry, A.J. Holden, D.J. Robbins, W. Stewart, Magnetism from conductors and enhanced nonlinear phenomena, IEEE Trans. Microw. Theory Tech. 47 (1999) 2075-2084. https://doi.org/10.1109/22.798002

[28] A. Pimenov, A. Loidl, P. Przyslupski, B. Dabrowski, Negative refraction in ferromagnet-superconductor superlattices, Phys. Rev. Lett. 95 (2005) 247000- 247009. https://doi.org/10.1103/PhysRevLett.95.247009

[29] A.M. Portis, Electrodynamics of high-temperature superconductors, World Scientific., Vol. 48, 1993, pp. 200-256 https://doi.org/10.1142/1867

[30] M. Ricci, N. Orloff, S.M. Anlage, Superconducting metamaterials, App.Phys.Lette. 87 (2005) 034100- 034102. https://doi.org/10.1063/1.1996844

[31] M.C. Ricci, S.M. Anlage, Single superconducting split-ring resonator electrodynamics, App. Phys. Lett. 88 (2006) 264100- 264102. https://doi.org/10.1063/1.2216931

[32] M.C. Ricci, H. Xu, R. Prozorov, A.P. Zhuravel, A.V. Ustinov, S.M. Anlage, Tunability of superconducting metamaterials, IEEE Trans. Appl. Supercond. 17 (2007) 918-921. https://doi.org/10.1109/TASC.2007.898535

[33] G. Romero, J.J. García-Ripoll, E. Solano, Microwave photon detector in circuit QED, Phys. Rev. Lett. 102 (2009) 173600-173602. https://doi.org/10.1103/PhysRevLett.102.173602

[34] R.A. Shelby, D. Smith, S. Nemat-Nasser, S. Schultz, Microwave transmission through a two-dimensional, isotropic, left-handed metamaterial, App. Phys. Lett. 78 (2001) 489-491. https://doi.org/10.1063/1.1343489

[35] R.W. Simon, R.B. Hammond, S.J. Berkowitz, B.A. Willemsen, Superconducting microwave filter systems for cellular telephone base stations, Procee. IEEE. 92 (2004) 1585-1596. https://doi.org/10.1109/JPROC.2004.833661

[36] I.I. Smolyaninov, V.N. Smolyaninova, Metamaterial superconductors, Nanophotonics. 7 (2018) 795-818. https://doi.org/10.1515/nanoph-2017-0115

[37] C.M. Soukoulis, J. Zhou, T. Koschny, M. Kafesaki, E.N. Economou, The science of negative index materials, J.Phys. Cond.Matter. 20 (2008) 304215- 304217. https://doi.org/10.1088/0953-8984/20/30/304217

[38] H. Tao, L.R. Chieffo, M.A. Brenckle, S.M. Siebert, M. Liu, A.C. Strikwerda, K. Fan, D.L. Kaplan, X. Zhang, R.D. Averitt, Metamaterials on paper as a sensing platform, Adv. Mater. 23 (2011) 3197-3201. https://doi.org/10.1002/adma.201100163

[39] M. Tinkham, Introduction to superconductivity, Courier Corporation, 2004

[40] G. Tsironis, N. Lazarides, I. Margaris, Wide-band tuneability, nonlinear transmission, and dynamic multistability in SQUID metamaterials, App. Phys. A. 117 (2014) 579-588. https://doi.org/10.1007/s00339-014-8706-7

[41] M. Wiltshire, J. Pendry, I. Young, D. Larkman, D. Gilderdale, J. Hajnal, Microstructured magnetic materials for RF flux guides in magnetic resonance imaging, Science. 291 (2001) 849-851. https://doi.org/10.1126/science.291.5505.849

[42] B. Wood, J. Pendry, Metamaterials at zero frequency, J. Phys.Cond. Matter. 19 (2007) 076200- 076208. https://doi.org/10.1088/0953-8984/19/7/076208

[43] P. Yang, C.M. Lieber, Nanorod-superconductor composites: a pathway to materials with high critical current densities, Science. 273 (1996) 1836-1840. https://doi.org/10.1126/science.273.5283.1836

Superconductors - Materials and Applications
Materials Research Forum LLC
Materials Research Foundations 132 (2022) 210-229
https://doi.org/10.21741/9781644902110-12

Chapter 12

Superconductors for Medical Applications

S.S. Ali[1*] and M. Zulqarnain[2]

[1]School of Physical Sciences, University of the Punjab, Lahore 54590, Pakistan

[2]Department of Physics, The University of Lahore, Lahore 54000, Pakistan

* shahbaz.sps@pu.edu.pk

Abstract

Superconductivity plays a vital role in advanced medical diagnostic as well as in treatment of cancer. Smaller sized superconducting cyclotrons are developing as efficient techniques with carbon ions and protons for external beam therapy. This equipment further gives benefits of less cost and due to smaller size it is far easier to handle. Nowadays, superconductivity has been used commercially in numerous applications in medical sciences including low-temperature superconducting (LTS) materials and high field magnets in magnetic resonance imaging (MRI), nuclear magnetic resonance (NMR), magnetic gene transfer, magnetic drug delivery system and cancer and internal hemorrhages detection. Almost all commercial medical systems based on superconductivity use LTS and the majority uses NbTi wires or superconducting quantum interference device (SQUID) made of LTS material.

Keywords

Magnetic Resonance Imaging (MRI), MRI based Food Inspection, Magnetic Gene Transfer, Magnetic Drug Delivery System (MDDS), Internal Hemorrhages Detection

Contents

1.   Introduction

In 1908, Kamerlingh Onnes's successfully manifested the liquefaction of Helium, and this led to the demonstration of zero electrical current resistivity in 1911 and hence the discovery of superconductivity has been presented to the world. It was conceived that the resistive limits of copper can be surmounted, no matter what the amount of current is, to realize superconductors without having any ohmic losses. Although progresses have been carried out, however, small Hc1 and narrow Jc of primary superconductors inhibited the practical applications during that time. In 1960s the practical superconducting materials including NbTi and Nb3Sn have been made to support sufficiently high magnetic fields and currents. Furthermore, enough long wires have been manufactured that can be used in magnet winding which truly demonstrated the applied superconductivity [1].

Currently, out of total Gross Domestic Product (GDP) of United States, a considerable part represents the health care spending and this is further growing swiftly [2]. Although the mentioned cost regarding US is the biggest one around the globe, however even in other countries the health care spendings are going up which rigorously affects the economies of these nations.

The conventional attitude of disease treatment is reactive in which people wait until someone is sick before treating the disease with costly processes. As a result, large annual amounts have been spent on treatments and smaller amounts were used for identification and diagnosis. This attitude needed to be rectified to inhibit the big spendings on healtcare system. Therefore, it has been proposed by medical experts that full body medical check-ups should be carried out at least twice in a year to get prior information if some disease is developing in the body before it gets worsen [3].

Superconducting magnets are quite important regarding application point of view. They are being used extensively in MRI/NMR systems. Apart from these two major applications, these magnets are advantageous in magnetic separation to stimulate particles with weaker magnetic susceptibility in a medium of even smaller or non magnetic nature. In addition, superconductors are efficiently implemented in the systems for the medical processess including food inspection, magnetic drug delivery, magnetic gene transfer and cancer and internal haemorrhages detection [4].

Superconductors - Materials and Applications　　　　　　　Materials Research Forum LLC
Materials Research Foundations 132 (2022) 210-229　　　　https://doi.org/10.21741/9781644902110-12

2.　Medical applications

2.1　Magnetic resonance imaging (MRI)

When a nucleus is positioned in an external magnetic field, it can absorb energy in the form of electromagnetic radiation of specific frequencies. This ability of nuclear magnetic resonance paves way towards the development of Magnetic Resonance Imaging (MRI) technology. The function of MRI is to provide the internal images of human body which further based on the distribution of hydrogen nuclei withing a particular organ under observation.

In modern days MRI has become a dynamic diagnostic technique which provides accurate detection and information without putting any harmful effects due to presence of magnetic field on living cells in human body. With the advancement in technological aspects, a number of screening processes like cancer treatment, neurosurgical planning and angiography are being done comfortably through MRI which mainly includes for cancer treatment, angiography and neurosurgical planning [5,6].

Key parameter to get high resolution images is the stronger magnetic fields, that is why, a number of research teams around the globe struggling to get even stronger magnetic fields in MRI systems with ensured safety parameters. This field strength will eventually leads to the improved observation and diagnosis and hence will be helpful to better understand body compositions followed by even better treatment procedures [7].

2.1.1　Quench protection design of MRI superconducting magnet

MRI superconducting magnet of 9.4 T strength has been reported by S. Chen et al. having 800 mm diameter bigger bore. A five coaxial solenoid coils superconducting magnet has been developed with NbTi Wire-in-Channel (WIC) conductor keeping Cu to non Cu ratio from 5 to 10. Maintaining a larger level of stored energy (138 MJ), the magnet was deisgned to function at smaller current of about 224.5 A. To prevent the damage of magnet a protection method has been adopted. A set up of series combination of resistors and diodes through sectioned parts has been established. Quench has been enhanced through an active trigger heater. A detailed analysis based on this protection scheme has been performed with quench simulations of voltages, currents and temperatures [8].

Formation of 9.4 T magnet for MRI system is shown in figure 1 illustrating MC1, MC2, MC3, MC4 and MC5 as the main coils and CC1 through CC4 as compensation coils. In table 1, other relevant specifications of MRI magnet having field strength of 9.4 T are tabulated. A field strength of 9.4 T is produced by the superconducting magnet maintainig a uniformity of 5×10^{-8} with a dimeter of 220 mm in the diameter spherical volume. Stored energy and operation current of the magnet are 138 MJ and 224.5 A respectively [9-11].

In simulation results, it has been assumed that initially the quench takes place in coil MC1_1. Figure 2 illustrates the variations in current, hot-spot temperature and voltage. After a time interval of 2.5 seconds the compensations coils and after 3 seconds all coils have been heated to quench. Considering the safety levels, highest voltage and current values of 277 V and 365 A respectively in CC2 and MC4_2, and the peak temperature of 98 K are all well below the safety levels during the quench. Owing to close thermal contact between heater strip and coil, the temperature of each heater strip is less and near to the hot-spot of coil. All these observations manifested a reliable model of quench protection for 9.4 T field strength magnet to be used in MRI systems [8].

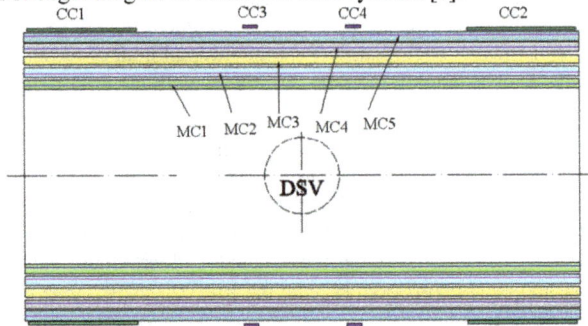

Figure 1.   Design of MRI magnet having 9.4 T strength [8].

Table 1.   Specifications of MRI magnet having field strength of 9.4 T

Coil	Inner diameter (cm)	Outer diameter (cm)	Length (cm)	Current density (A/cm^2)
MC1	91	101.0	300	2155.5
MC2	103.8	113.8	300	2405.85
MC3	116.6	126.1	300	2429.3
MC4	129.1	138.9	300	3163.25
MC5	142.1	150.9	300	6964.07
CC1	152	155.2	59.3	12,130
CC2	152	155.2	59.3	12,130
CC3	155	157.9	8.1	−12,136
CC4	155	157.9	8.1	−12,136

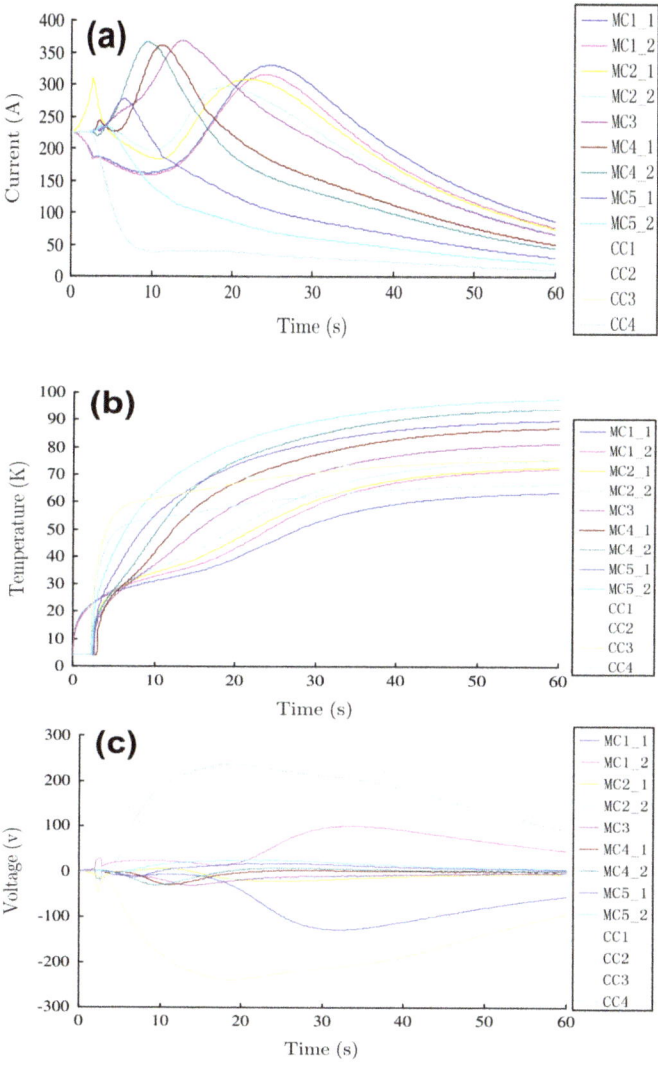

Figure 2.   Current, hot-spot temperature and voltage curves during quench [8].

Superconductors - Materials and Applications   Materials Research Forum LLC
Materials Research Foundations 132 (2022) 210-229   https://doi.org/10.21741/9781644902110-12

2.1.2 Open MRI superconducting magnet

In magnetic resonance imaging (MRI) systems the crucial part is the magnet, which ensures the creation of high quality images through its exposure on body organs [12,13]. Uniformity of magnetic and stray felds should be controlled. In an open MRI system there is significant gap between the two poles of superconducting magnet [14–17]. This gap causes considerable difficulties in handling the uniformity of magnetic field, the balance of electromagnetic force and the performance of superconducting wire as well. Apart from all these hindrances, there are a number of benefits of open MRI systems as compared to conventional systems. For instance, it significantly reduces the claustrophobia feeling of patient due to bigger bore for patient and comparatively smaller length [18]. It also gives liberty to move the patient for better adjustment againt exposure ensuring better images.

The initial approximation of a magnet design is proceeded through linear optimization principle which is shown in figure 3 as a meshed magnet space. The current estimation of existing techniques are shown in figure 3 (b) [20-25]. Without cross sectional area, for each element the thick solenoid is designed as current loop [26]. The unidentified currents of the rectangular are found through linear optimization. A concentrated current is taken in to account in this technique to show the distributed current and avoiding the dimensional content. Therefore, it is to be put on a thick mesh on certain difficult designs, for instance the magnet with extremly small size. Fig. 3(c) represents a design procedure, in which (z_1, z_2, r_1, r_2) represents coordinate of the solenoid [19].

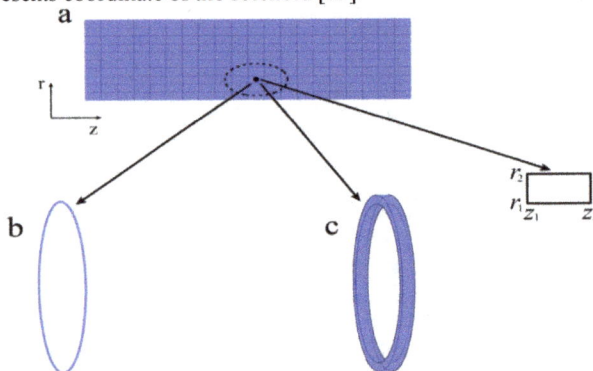

Figure 3.   Magnet design linear optimization: (a) meshed magnet space (b) filament-loop approximation (c) use of thick solenoid for magnetic field calculation [19].

For a planar magnet fig. 4(a) shows the magnet-space mesh of the cross-section, the structure supports are indicated by slashes. Due to regular design, just a quarter has been taken into account in design procedure [Fig. 4(b)]. The diameter and height of magnet coil in meters are 2.4 and 2.0 repsectively. These dimensions are suitable for 2.0 T magnet scheme [27] and open MRI set up of 1.0 T. The magnet scheme is between two open systems for 1.5 T with exterior diameter bigger than 1.2 T and lesser than 2.0 T system [16,27]. As compared to 0.7 T open MRI design, the open bore has larger spacing of 0.72 m and diameter of 1.5 m [21].

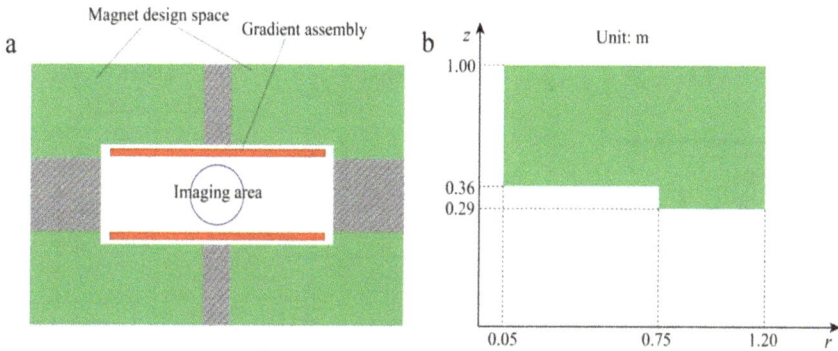

Figure 4.   The mesh structure of the magnet-space: (a) the rectangular mesh for a full magnet model (b) a quarter of the full mesh model [19].

2.1.3 MRI food inspection system

Nuclear magnetic resonance (NMR) and magnetic resonance imaging (MRI) with ultra low field (ULF) have attracted considerable attention due to its proposed compact size, easiness in handling and being inexpensive as compared to existing NMR/MRI systems [28-31]. SQUID can be employed as a detector of NMR signals at such a low frequency range due to efficient sensitivity without being influenced from frequency. Low temperature superconductor (LTS) SQUID has smaller magnetic noise as compared to that of HTS, however, still HTS SQUID has been used in a number of applications because of its merits including lower cost, compact size, better portability and user friendliness as well [32-35].

For the detection of contaminent in food and drink, utilizing an HTS-SQUID, *Y. Hatsukade et al.* developed an ULF nuclear magnetic resonance (NMR) and magnetic resonance imaging (MRI) system. In detection procedure the protons in water samples have been polarized through a 1.1 T magnet. NMR signals have been measured for pure water and the one having impurities of aluminum, stainless steel and glass. Contaminant detection has been performed based on the fact that in the presence of SUS304 ball, the signal intensity was much smaller as compared to the intensity in the absence of contaminant owing to the remnant field. Non magnetic contaminants have been detected through one-dimensional (1D) MRIs. A change in position of contaminant causes corresponding change in 1D MRI spectra. Theses observations show the effciency of the system for identification of contaminations in food items [36].

The ULF-NMR/MRI setup developed by Y. Hatsukade et al. is presented in Fig. 5. The major components of the system are cryostat, Helmholtz-type measurement coil, HTS-rf-SQUID, SQUID electronics, sample transfer apparatus, function generator and spectrum analyzer.

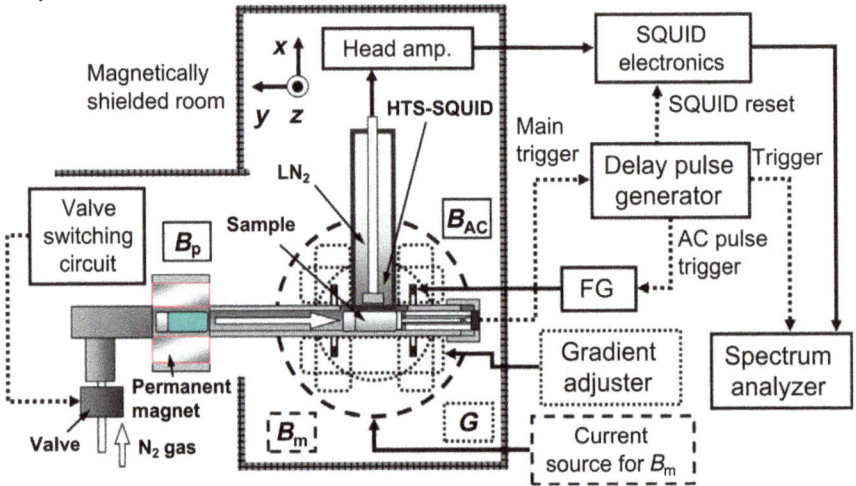

Figure 5.   HTS-SQUID ULF-NMR/MRI system manifested schemtically [36].

For measurements, a magnetic field of 45 µT and 56 nT/cm gradient field has been applied too. Fig. 6 (a-d) illustrates the MRI spectra of pure and contaminated water samples, where contaminants are (a) ceramic, (b) aluminum, (c) glass and (d) nylon. The MRI spectra of contaminated samples is altered around the frequency of about 1914 Hz unlike the analysis for pure water. The particular mentioned frequency corresponds to the location of the contaminants. The presence of contaminants influences the distribution of protons of water [36].

2.2   Magnetic gene transfer

Introduction of tumor supression genes in carcinomatous lesion makes gene therapy an efficient tool for cancer treatment. In this technique gene carriers are inoculated in the body part affected by cancer, however the uptake ability is declined through DNA carriers diffusion. This depreciated uptake efficiency of cell after wide distribution and diffusion of DNA carriers is a problem to be addressed. One direct solution to this problem is to increase the contact probablity of DNA with the tumor cell. This could be possible by putting some physical force on DNA carriers to get them concentrated in a localized region of interest and this in turn improve the effectivness of gene transfer into the affected tissues.

Such investigation has been carried out by K. Nakagawa et al. considering the transmission of ferromagnetic DNA carriers. This diffusion has been realized in the presence of magnetic field with higher intensity produced by HTS magnet [37, 38].

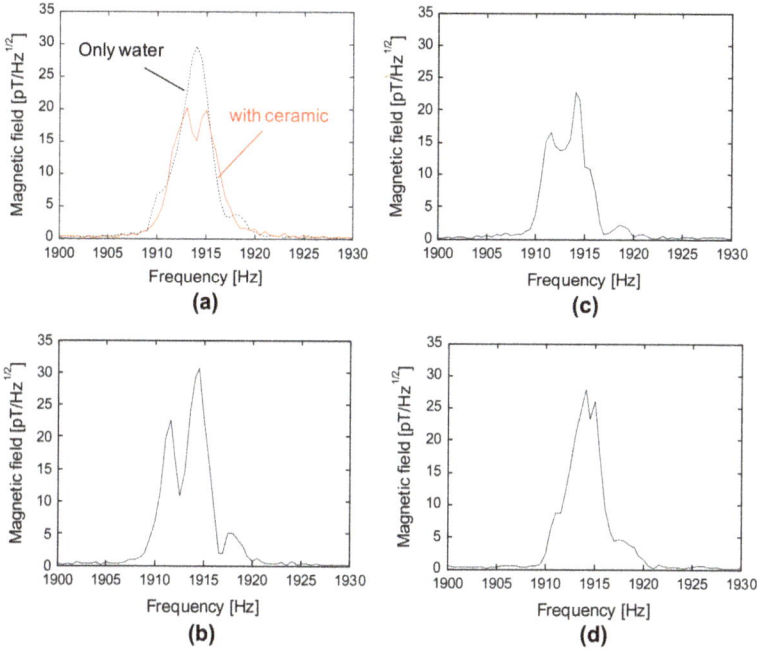

Figure 6. Observations of 1D-MRI for (a) Just water, and water + ceramic (b) Water + aluminum ball (c) Water + glass ball (d) Water + nylon ball. Ball diameter = 5 mm [36].

Magnetic gene transfer has been analyzed using the ferromagnetic DNA carriers (Magnetite complexes). Supressed particle diffusion has been analyzed up to 10 mm through simulated results at the cost of magnetic field. The outcomes of investigations revealed enhanced gene expression through magnetic field.

Aligning the target region with the central axis of magnet, the insertion of the particles was made at the bottom of the said region and based on this scheme the diffusion was calculated (Fig. 7). After 30 minutes of inoculation, the calculated results of horizontal distribution of particles are shown in Fig. 8. The particle distribution rate in the range of 1 cm was found 27% in the absence of magnet, however, 44%, 59% and 73% at 50 mm, 30 mm and 20 mm respectively above the magnet. As compared to the situation of without having a magnet, the concentration of carriers has been found 2.7 times larger in the case when the magnet was placed at 20 mm from the target [37].

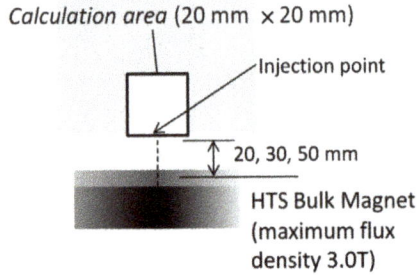

Figure 7.   *A representation of simulation model.*

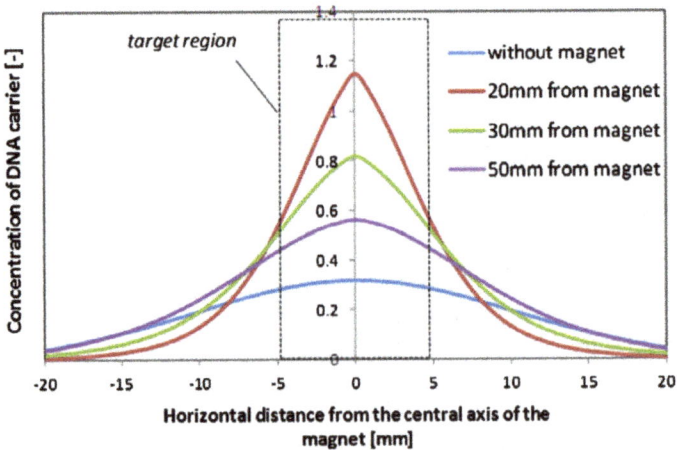

Figure 8.   *The estimated horizontal distribution of particle after 30 min of insertion.*

2.3   Magnetic drug delivery system

A scheme which is being used to control the distribution and circulation of the drug dose in the body as well as at the target (affected part or organ) is known as drug delivery system (DDS). The system can minimize the unsafe effects and enhances the medicinal effects. In order to ensure more efficiency than the DDS, a method called magnetic drug delivery system (MDDS) has been employed in which magnetic forces are used to regulate the ferromagnetic drug to the target within a human body.

M. Chuzawa et al. reported the MDDS method to ensure the ferromegnetic drug so that its dose may concentrate at localized regions with the application of strong magnetic field provided by HTS magnets [39-46]. They attempted to alter the magnetic field, so that a uniform and intense field can be maintained in body to gather ferromagnetic drugs.

Figure 9 shows Magnet-rotating MDDS analyzing the confined concentration in a liver diseased part having around 10 mm of diameter. The spreading of magnetic field is presented in Fig. 10. The simulation has been performed while keeping a magnet circling around the capillary vessel. The coordinates have been selected such that its location within the vessel's inlet remains exactly at the center. The three axes have been adjusted as: x along the blood flow, y is vertically downword and z is at 90° to both x and y axes [Fig. 10].

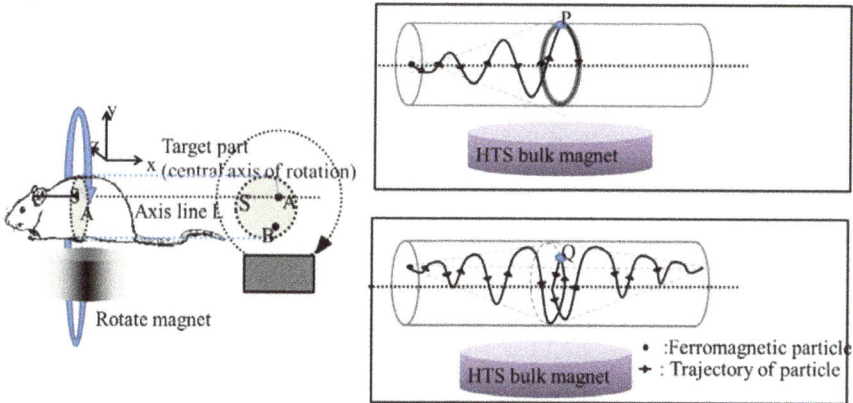

Figure 9.   Magnet-rotating MDDS illustration for a rat [47].

The kinetics of ferromagnetic drugs inside the body have been analyzed through simulation in order to realize applied MDDS. It has been manifested that, it is possible to concentrate the ferromagnetic particles at the target region. Such analyses can be carried out in future investigations to further mature the procedure for the accumulation of drug particles at certain body parts. Furthermore, for the accuracy of the simulation outcomes, the model experiment can be performed by using a model organ [47].

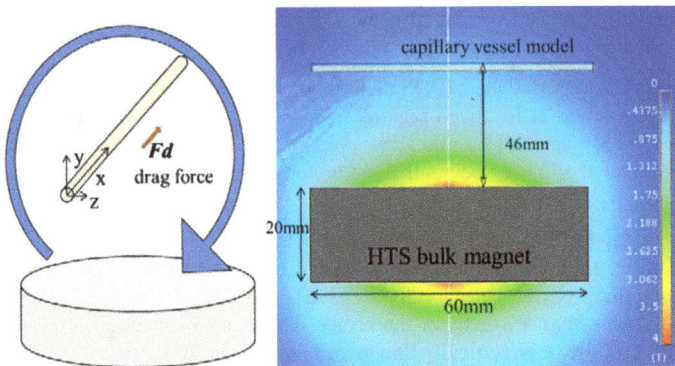

Figure 10. In a Magnet-rotating MDDS, the distribution magnetic field of HTS bulk magnet.

2.4 Cancer and internal hemorrhage detection

Lorentz Force Electrical Impedance Tomography (LFEIT) is an advanced, portable, and easy to operate device which can be used for the detection of both cancer and internal hemorrhage [48-50]. As the value of the electrical impedance alters significantly in case of soft tissues as well as in pathological state so that is why LFEIT possesses excellent capability to provide pathological and physiological situation of body parts whereas ultrasonic imaging technology is unable to detect such affected tissues [51]. Apart from carcinomas, body tissues in the situation of hemorrhage or ischemia displays different permittivity and conductivity of body fluid and blood as compared to other soft tissues i.e. both exhibit different electrical properties [52]. An LFEIT arrangement is shown schematically in figure 11. LFEIT works on the principle of measurement of electrical signals emerging on the propagation of an ultrasound wave through condutive region, and the region is subjected perpendicularly to a magnetic field. This electrical signal has a direct relation with magnetic field intensity and ultrasound wave pressure [48].

A model of superconducting Halbach Array (HA) magnet consisting of 8 coils (90° phase change for each coil) and 12 coils (60° phase change for each coil) was presented by *B. Shen et al.* as shown in Fig. 12. Partial Differential Equation (PDE) model with 2D *H*-formulation, COMSOL Multiphysics has been utilized for the modelling of superconducting magnet. Multiple tapes having sheets of layered conductor have been illustrated by continuous area bulk approximation. These have been employed to enhance the speed of model simulation [53].

Fig. 13 (a,b) shows the loop for superconducting Halbach Array designs consisting of 8 and 12 HTS coils respectively. Coils' stacks are denoted via bulk approximation. As manifested in Fig. 13, air is in the centre of the ring. Each coil contains a supply of 120 A direct current. Multiple coils can be used by Halbach Array configuration keeping the total amount of superconductors same. This can be achieved through reduction of size of each coil and enhancing amount of coils from 8 to 12 [53].

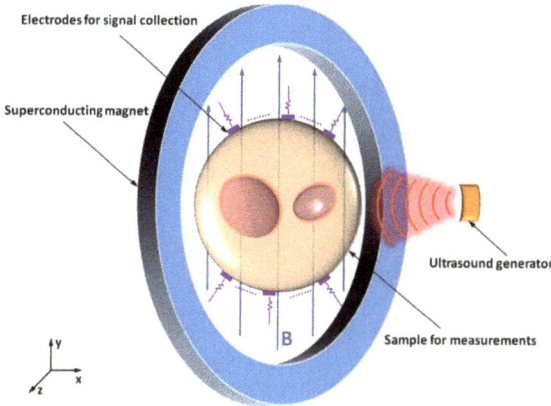

Figure 11.   Schematic representation of superconducting Lorentz Force Electrical Impedance Tomography (LFEIT).

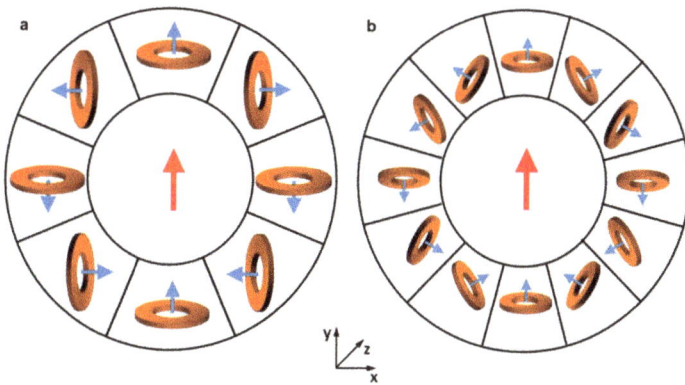

Figure 12.   Superconducting Halbach Array electromagnet with, (a) 8 and (b) 12, HTS coils.

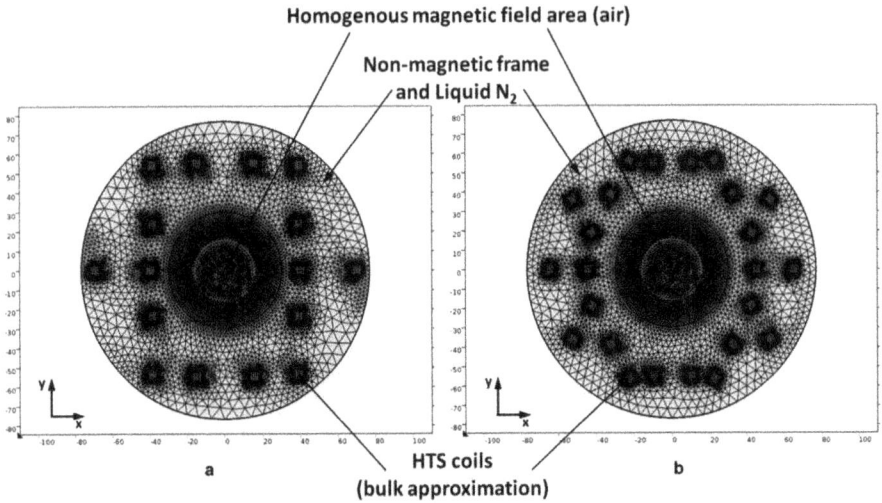

Figure 13.   Halbach Array designs having, (a) 8 and (b) 12, HTS coils.

Conclusions

Over the decades after their discovery, superconductors have been emerged as powerful and efficient materials for a number of commercial applications specifically including magnetic resonance imaging (MRI), nuclear magnetic resonance spectroscopy (NMR), magnetic drug delivery systems, magnetic gene transfer mechanism and their use in cancer treatments. With the passage of time superconductors find even more and more applications in field of medical devices. A number of research groups have investigated and still working to achieve critical temperatures as high as possible to get it closer and closer to room temperature. This ofcourse will let the industry to get rid of helium and nitrogen based cryogenics for low and high temperature superconductors respectively. The over all high cost of HTS based applications is still the dominent factor to limit the widespread of technology. Supposedly, even if all the technical problems yet existing with HTS are resolved, still HTS based systems are quite expensive which inhibits their applicability in a number of poor of even developing countries. It is believed that prevalent use of HTS in medical sector would be possible if some new families of superconductors will be developed. Such room temperature superconductors will not depend on cryogenics making them less in cost, squeezed in size and more feasible regarding maintenance.

References

[1] P. F. Dahl, Superconductivity: its historical roots and development from mercury to the ceramic oxides, American Institute of Physics, New York,

1992, pp 261-264.

[2] From data compiled by the U.S. Centers for Medicare & Medicaid Services. 7500 Security Boulevard, Baltimore, MD 21244, USA.

[3] M. O'Donnell, Magnetic nanoparticles as contrast agents for molecular imaging in medicine, Phys. C: Supercond. Appl. 548 (2018) 103-106. https://doi.org/10.1016/j.physc.2018.02.031

[4] A. Bergen, R. Andersen, M. Bauer, H. Boy, M. Ter Brake, P. Brutsaert, C. Buhrer, M. Dhalle, J. Hansen, H. Ten Kate and J. Kellers, Design and in-field testing of the world's first ReBCO rotor for a 3.6 MW wind generator, Supercond. Sci. Technol. 32 (2019) 125006. https://doi.org/10.1088/1361-6668/ab48d6

[5] M. Parizh, Y. Lvovsky and M. Sumption, Conductors for commercial MRI magnets beyond NbTi: requirements and challenges, Supercond. Sci. Technol. 30 (2017) 014007 https://doi.org/10.1088/0953-2048/30/1/014007

[6] Z. Wang, J. M. van Oort, and M. X. Zou, Development of superconducting magnet for high-field MR systems in China. Phys. C: Supercond. Appl. 482 (2012) 80-86. https://doi.org/10.1016/j.physc.2012.04.027

[7] T. C. Cosmus, and M. Parizh, Advances in whole-body MRI magnets, IEEE Trans. Appl.Supercond. 21 (2010) 2104-2109. https://doi.org/10.1109/TASC.2010.2084981

[8] S. Chen, Y. Li, Y. Dai, Y. Lei and L. Yan, Quench protection design of a 9.4 T whole-body MRI superconducting magnet, Phys. C: Supercond. Appl. 497 (2014) 49-53 https://doi.org/10.1016/j.physc.2013.11.001

[9] Y. Dai, Q. Wang, C. Wang, L. Li, H. Wang, Z. Ni, S. Song, S. Chen, B. Zhao, H. Wang and Y. Li, Structural design of a 9.4 T whole-body MRI superconducting magnet, IEEE Trans. Appl. Supercond. 22 (2012) 4900404. https://doi.org/10.1109/TASC.2012.2184509

[10] Q. Wang, High field superconducting magnet: Science technology and applications, Prog. Phys. 33 (2013) 1-23.

[11] Q. Wang, Y. Dai, B. Zhao, S. Song, C. Wang, L. Li, J. Cheng, S. Chen, H. Wang, Z. Ni and Y. Li, A superconducting magnet system for whole-body metabolism imaging, IEEE Trans. Appl. Supercond. 22 (2011) 4900905. https://doi.org/10.1109/TASC.2011.2175888

[12] B. M. Dale, M. A. Brown, R. C. Semelka, MRI: Basic Principles and Applications, Wiley-Blackwell/John Wiley & Sons, Hoboken, N. J. 2010.

[13] R. H. Hashemi, W. G. Bradley and C. J. Lisanti, MRI: the basics: The Basics. Lippincott Williams & Wilkins, Philadelphia, 2012.

[14] J. F. Schenck, F. A. Jolesz, P. B. Roemer, H. E. Cline, W. E. Lorensen, R. Kikinis, S. G. Silverman, C. J. Hardy, W. D. Barber, E. T. Laskaris and B. Dorri,

Superconducting open-configuration MR imaging system for image-guided therapy,
Radiology 195 (1995) 805-814. https://doi.org/10.1148/radiology.195.3.7754014

[15] T. Tadic and B. G. Fallone, Design and optimization of superconducting MRI
magnet systems with magnetic materials. IEEE Trans. Appl. Supercond. 22 (2012)
4400107. https://doi.org/10.1109/TASC.2012.2183871

[16] Y. Lvovsky, E. W. Stautner and T. Zhang, Novel technologies and configurations of
superconducting magnets for MRI, Supercond. Sci. Technol. 26 (2013) 093001.
https://doi.org/10.1088/0953-2048/26/9/093001

[17] S. Kakugawa, N. Hino, A. Komura, M. Kitamura, H. Takeshima, T. Yatsuo and H.
Tazaki, Three-dimensional optimization of correction iron pieces for open high field
MRI system, IEEE Trans. Appl. Supercond. 14 (2004) 1624-1627.
https://doi.org/10.1109/TASC.2004.831019

[18] M. Dewey, T. Schink and C. F. Dewey, Claustrophobia during magnetic resonance
imaging: cohort study in over 55,000 patients. J. Magn. Reson. Imaging 26 (2007)
1322-1327. https://doi.org/10.1002/jmri.21147

[19] Y. Wang, Q. Wang, L. Wang, H. Qu, Y. Liu and F. Liu, Electromagnetic design of a
1.5 T open MRI superconducting magnet, Phys. C: Supercond. Appl. 570 (2020)
1353602. https://doi.org/10.1016/j.physc.2020.1353602

[20] Z. Ni, Q. Wang, F. Liu, L. Yan, A homogeneous superconducting magnet design
using a hybrid optimization algorithm, Meas. Sci. Technol. 24 (2013) 125402.
https://doi.org/10.1088/0957-0233/24/12/125402

[21] Z. Ni, G. Hu, L. Li, G. Yu, Q. Wang, L. Yan, Globally optimal algorithm for design
of 0.7 T actively shielded whole-body open mri superconducting magnet system, IEEE
Trans. Appl. Supercond. 23 (2013) 4401104.
https://doi.org/10.1109/TASC.2012.2231720

[22] Q. M. Tieng, V. Vegh, I. M. Brereton, Globally optimal superconducting magnets
part II: symmetric MSE coil arrangement, J. Magn. Reson. 196 (2009) 7-11.
https://doi.org/10.1016/j.jmr.2008.09.023

[23] Q. M. Tieng, V. Vegh, I. M. Brereton, Globally optimal superconducting magnets
part I: Minimum stored energy (MSE) current density map, J. Magn. Reson. 196
(2009) 1-6. https://doi.org/10.1016/j.jmr.2008.09.018

[24] Z. Ni, G. Hu, Q. Wang, L. Yan, Globally optimal superconducting homogeneous
magnet design for an asymmetric 3.0 T head MRI scanner, IEEE Trans. Appl.
Supercond. 24 (2013) 1-5. https://doi.org/10.1109/TASC.2013.2283653

[25] X. Hao, S. M. Conolly, G. C. Scott, A. Macovski, Homogeneous magnet design
using linear programming, IEEE Trans. Magn. 36 (2000) 476-483.
https://doi.org/10.1109/20.825817

[26] Z. Ni, Q. Wang, F. Liu, L. Yan, A homogeneous superconducting magnet design using a hybrid optimization algorithm, Meas. Sci. Technol. 24 (2013) 125402. https://doi.org/10.1088/0957-0233/24/12/125402

[27] A. Borceto, D. Damiani, A. Viale, F. Bertora, R. Marabotto, Engineering design of a special purpose functional magnetic resonance scanner magnet, IEEE Trans. Appl. Supercond. 23 (2012) 4400205. https://doi.org/10.1109/TASC.2012.2234811

[28] R. McDermott, A.H. Trabesinger, M. Muck, E.L. Hahn, A. Pines and J. Clarke, Liquid state NMR and scalar coupling in microtesla magnetic fields, Science 295 (2002) 2247-2249. https://doi.org/10.1126/science.1069280

[29] R. McDermott, N. Kelso, S.-K. Lee, M. Moble, M. Muck, W. Myers, B. T. Haken, H. C. Seton, A. H. Trabesinger, A. Pines, J. Clarke, SQUID-detected magnetic resonance imaging in microtesla magnetic fields, J. Low Temp. Phys. 135 (2004) 793. https://doi.org/10.1023/B:JOLT.0000029519.09286.c5

[30] A. N. Matlachov, P. L. Volegov, M. A. Espy, J. S. George and R. H. Kraus Jr, SQUID detected NMR in microtesla magnetic fields, J. Magn. Reson. 170 (2004) 1-7. https://doi.org/10.1016/j.jmr.2004.05.015

[31] P. Volegov, A. N. Matlachov, M. A. Espy, J. S. George and R. H. Kraus Jr., Simultaneous magnetoencephalography and SQUID detected nuclear MR in microtesla magnetic fields, Magn. Reson.Med.52 (2004) 467-470. https://doi.org/10.1002/mrm.20193

[32] L. Q. Qiu, Y. Zhang, H. J. Krause, A. I. Braginski, A. Offenhäusser, J. Magn. Reson. 196 (2009) 101-104 https://doi.org/10.1016/j.jmr.2008.09.009

[33] L. Qiu, Y. Zhang, H. J. Krause, A. I. Braginski and A. Offenhausser, Low-field NMR measurement procedure when SQUID detection is used, J. Magn. Reson.196 (2009) 101-104. https://doi.org/10.1016/j.jmr.2008.09.009

[34] S. Y. Yang, K. W. Lin, J. J. Chieh, C. C. Yang, H. E. Horng, S. H. Liao, H. H. Chen, C. Y.Hong and H.C. Yang, Step-Edge High-Tc SQUID Magnetometer for Low-Field NMR Detection, IEEE Trans. Appl. Supercond. 21 (2011) 534-537. https://doi.org/10.1109/TASC.2011.2104353

[35] S. Fukumoto, M. Hayashi, Y. Katsu, M. Suzuki, R. Morita, Y. Naganuma, Y. Hatsukade, S. Tanaka, O. Snigirev, Liquid-State Nuclear Magnetic Resonance Measurements for Imaging Using HTS-rf-SQUID in Ultra-Low Field, IEEE Trans. Appl. Supercond. 21 (2011) 522 https://doi.org/10.1109/TASC.2011.2106474

[36] Y. Hatsukade, S. Tsunaki, M. Yamamoto, T. Abe, J. Hatta and S. Tanaka, Feasibility study of contaminant detection for food with ULF-NMR/MRI system using HTS-SQUID", Phys. C: Supercond. 494 (2013) 199-202. https://doi.org/10.1016/j.physc.2013.04.004

[37] K. Nakagawa, Y. Ohaku, J. Tamada, F. Mishima, Y. Akiyama, M. Kiomy Osako, H. Nakagami and S. Nishijima, Study on magnetic gene transfer using HTS bulk magnet, Phys. C: Supercond. 494 (2013) 262-264 https://doi.org/10.1016/j.physc.2013.04.072

[38] J. P. Fortin-Ripoche, M. S. Martina, F. Gazeau, C. Menager, C. Wilhelm, J. C. Bacri, S. Lesieur and O. Clement, Magnetic targeting of magnetoliposomes to solid tumors with MR imaging monitoring in mice: feasibility, Radiology 239 (2006) 415-424 https://doi.org/10.1148/radiol.2392042110

[39] S. Takeda, F. Mishima, B. Terazono, Y. Izumi, S. Nishijima, Development of Magnetic Force-Assisted New Gene Transfer System Using Biopolymer-Coated Ferromagnetic Nanoparticles, IEEE Trans. Appl. Supercond. 16 (2006) 1543. https://doi.org/10.1109/TASC.2005.869695

[40] F. Mishima, S. Fujimoto, S. Takeda, Y. Izumi, S. Nishijima, Development of control system for magnetically targeted drug delivery, J. Magn. Magn. Mater. 310 (2007) 2883. https://doi.org/10.1016/j.jmmm.2006.11.124

[41] F. Mishima, S. I. Takeda, Y. Izumi and S. Nishijima, Development of magnetic field control for magnetically targeted drug delivery system using a superconducting magnet, IEEE Trans. Appl. Supercond. 17 (2007) 2303-2306. https://doi.org/10.1109/TASC.2007.898413

[42] S. Nishijima, F. Mishima, T. Terada and S. Takeda, A study on magnetically targeted drug delivery system using superconducting magnet, Phys. C: Supercond. Appl. 463 (2007) 1311-1314. https://doi.org/10.1016/j.physc.2007.03.493

[43] Y. Yoshida, S. Fukui, S. Fujimoto, F. Mishima, S. Takeda, Y. Izumi, S. Ohtani, Y. Fujitani and S. Nishijima, Ex vivo investigation of magnetically targeted drug delivery system, J. Magn. Magn. Mater. 310 (2007) 2880-2882. https://doi.org/10.1016/j.jmmm.2006.11.123

[44] F. Mishima, S. Fujimoto, S. Takeda, Y. Izumi and S. Nishijima, Development of control system for magnetically targeted drug delivery, J. Magn. Magn. Mater. 310 (2007) 2883-2885. https://doi.org/10.1016/j.jmmm.2006.11.124

[45] T. Terada, S. Fukui, F. Mishima, Y. Akiyama, Y. Izumi and S. Nishijima, Development of magnetic drug delivery system using HTS bulk magnet, Phys. C: Supercond. 468 (2008) 2133-2136. https://doi.org/10.1016/j.physc.2008.05.145

[46] M. Chuzawa, F. Mishima, Y. Akiyama, S. Nishijima, Drug accumulation by means of noninvasive magnetic drug delivery system, Phys. C: Supercond. Appl. 471 (2011) 1538-1542. https://doi.org/10.1016/j.physc.2011.05.233

[47] M. Chuzawa, F. Mishima, Y. Akiyama and S. Nishijima, Precise control of the drug kinetics by means of non-invasive magnetic drug delivery system, Phys. C: Supercond. 484 (2013) 120-124. https://doi.org/10.1016/j.physc.2012.03.070

[48] P. Grasland-Mongrain, J. M. Mari, J. Y. Chapelon and C. Lafon, Lorentz force electrical impedance tomography, Irbm 34 (2013) 357-360. https://doi.org/10.1016/j.irbm.2013.08.002

[49] Y. Xu and B. He, Magnetoacoustic tomography with magnetic induction (MAT-MI), Phys. Med. Biol. 50 (2005) 5175. https://doi.org/10.1088/0031-9155/50/21/015

[50] H. Wen, J. Shah and R.S. Balaban, Hall effect imaging, IEEE Trans. Biomed. Eng. 45 (1998) 119-124. https://doi.org/10.1109/10.650364

[51] C. Gabriel, S. Gabriel and Y. E. Corthout, The dielectric properties of biological tissues: Literature survey, Phys. Med. Biol. 41 (1996) 2231. https://doi.org/10.1088/0031-9155/41/11/001

[52] H. P. Schwan and K. R. Foster, RF-field interactions with biological systems: electrical properties and biophysical mechanisms, Proc. IEEE 68 (1980) 104-113. https://doi.org/10.1109/PROC.1980.11589

[53] B. Shen, L. Fu, J. Geng, X. Zhang, H. Zhang, Q. Dong, C. Li, J. Li and T. A. Coombs, Design and simulation of superconducting lorentz force electrical Impedance tomography (LFEIT), Phys. C: Supercond. Appl. 524 (2016) 5-12. https://doi.org/10.1016/j.physc.2016.02.023

Superconductors - Materials and Applications
Materials Research Foundations 132 (2022) 230-255

Materials Research Forum LLC
https://doi.org/10.21741/9781644902110-13

Chapter 13

Superconductors for Magnetic Imaging Resonance Applications

Ali Raza[1] and S.S. Ali[2*]

[1]Department of Physics, The University of Lahore, Lahore 54000, Pakistan

[2]School of Physical Sciences, University of the Punjab, Lahore 54590, Pakistan

* shahbaz.sps@pu.edu.pk

Abstract

Magnetic Resonance Imaging (MRI), a medical imaging technique is being widely used in diagnostic procedures. This technique is based on radiology and generates clear pictures of the anatomy and other physiological processes in the human body. A strong magnetic field is the key parameter for better resolution images provided it does not damage healthy tissues. The fundamental principle behind MRI is the response of the atomic nucleus against the magnetic field which is followed by the image creation based on the distribution of hydrogen nuclei in a specific organ under observation. Stronger magnetic fields produce improved images with higher resolutions, that is why several research groups around the globe are working to develop such MRI systems without putting harmful effects on the exposed organs.

Keywords

MgB_2 Superconductors, Bi-2223, HTS Magnets, REBCO, Iron-Based Superconductors

Contents

1. Introduction

A substance that possesses no electrical resistance and no magnetic forces can penetrate through it is known as a superconductor. When certain substances are cooled to a certain temperature, superconductivity occurs, in which all electrical resistance vanishes.

Superconductivity is often only possible at extremely low temperatures. In everyday life, there are many applications of superconductors including MRI machines and maglev trains in which the whole train is lifted from the track to get rid of friction. Superconductors that can work at higher temperatures are now being developed by researchers, which might change energy storage and transportation. It is quite important nowadays to develop such materials which manifest superconductivity under normal conditions, unlike the already existing ones which are superconductors at higher pressures or low temperatures. The following stage is to formulate a hypothesis that explains how the new superconductors function and predicts their characteristics.

Low-temperature superconductors (LTS) and high-temperature superconductors (HTS), also known as conventional and unconventional superconductors, are two types of superconductors. HTS uses different microscopic approaches to attain zero resistance, whereas LTS uses the Bardeen-Cooper-Schrieffer (BCS) theory which tells about the formation of Cooper pairs. One of the main unanswered mysteries in contemporary physics is the origins of HTS. Being convenient to develop/discover, the majority of historical superconductivity research has been focused on low-temperature superconductors. Almost all superconducting applications need low-temperature superconductors.

In today's world, high-temperature superconductors, on the other hand, is a thriving and fascinating subject of study. Anything that operates as a superconductor beyond 70 K is referred to as high-temperature superconductor (HTS). Much though that temperature is still rather cold, still it is favorable because liquid nitrogen can be utilized to get this much temperature making this category of superconductors quite inexpensive as compared to liquid helium.

2. History of superconductor materials for MRI

Several superconducting materials have been found after the famous date of 1911, the list of which and their history are outside this text's scope. NbTi, Nb_3Sn, and MgB_2 have emerged as the materials of choice for making superconducting magnets, with NbTi acting as the true workhorse at a fraction of the cost of Nb_3Sn and MgB_2. The use of superconductors took advantage of the fact that copper was expensive and required a lot more power, making the related heat difficult to manage. High energy physicists pushed for the usage of superconductors.

In 1960, the Rutherford Appleton Laboratory in the United Kingdom created the first superconducting accelerators using copper-coated NbTi, followed by American scientists in 1967. In the same year that commercial NbTi was produced, the world's first 10T magnet was built by General Electric (GE). Even though this was seen as a triumph, the business component was a failure, since Bell Labs only paid GE \$75 k for a magnet that cost \$200 k to develop [1]. Intermagnetics General Corporation and Oxford Instruments emerged as champions of superconductor commercialization. In 1967, OI demonstrated a 10 T magnet with a 30 mK dilution refrigerator. Yet, magnetic resonance imaging (MRI) and, to a lesser degree, nuclear magnetic resonance (NMR) has extended the application of superconductors.

To date, Nuclear Magnetic Resonance and Magnetic Resonance Imaging remain the most profitable applications of superconductivity. In more than 75 percent of MRI magnets, superconducting wires are employed. Over 2,500 superconducting MRI systems are sold every year, and over 75,000 people are scanned every day. There were around 26,500 superconducting MRI systems deployed till 2008, compared to 14,600 in 2002 [2].

The most successful commercial uses of superconductivity to date have been MRI (and NMR) magnets. In the 1980s, the widespread application of these magnets began in earnest. By this time, a solution had been discovered to each of the distinct and common design and operation issues that these magnets faced, allowing them to become commercially viable products. These concerns are:

1. The superconductor itself, notably NbTi technology, which was, is, and will continue to be a mainstay of both magnets for ten years ahead

2. Ways for making superconducting joints

3. Heat transfer equilibrium

4. Protection measures

NbTi has been the superconductor of choice for every MRI magnet manufacturer for a long time. Oxford Instruments and other manufacturers used an adiabatic design approach to make the magnet commercially feasible from the start, which means that winding has no local cooling by liquid helium. These adiabatic magnets having narrow 'energy margins' were affected from early quenches, which were mostly caused by mechanical disturbances [3], since they were not locally cooled. These efficient NbTi magnets are no longer as susceptible to mechanical disturbances as they formerly were, although they are nevertheless vulnerable at times. The development of HTS revolutionized the concept in superconducting magnet technology, altering critical design parameters such as stability [4].

2.1   Liquid helium free SN2 high-temperature fuperconductor magnet

The notion of combining a Liquid Helium free high-temperature superconductor magnet and solid cryogen, notably solid nitrogen (SN2), as a superconducting 'permanent' magnet was initially presented in 1997 [5]. A superconducting 'permanent' magnet operating in persistent mode could easily provide a field well above that of a conventional permanent magnet, i.e., > 0.5 T, for an extended duration.

2.2   Bismuth strontium calcium copper oxide (Bi2223): First SN2–HTS magnet

In 2001, the first SN2–HTS proof-of-concept magnet was completed [6,7]. Six double-pancake (DP) coils are connected in series and wrapped in Bi2223 tape to make the high-temperature superconductor magnet. A schematic representation of the device is shown in Figure 1. In this experiment, LHe was utilized as the cooling source to solidify the liquid nitrogen, with the main goal of demonstrating the utility of solid nitrogen thermal mass (LN2). Figure 2 shows how LHe was utilized to reduce the temperature from 77 to 20 degrees Celsius. A copper chamber encloses the magnet and a persistent-current switch (PCS) that is hung in an aluminum cryostat. A couple of Cu tubes are arranged in the form of an annulus to make the cold body's vertical walls. During warm-up, a channel is provided by one tube to fill the container with LN2 and eject N2, meanwhile, the other two tubes offer a way for the circulation of liquid helium via the helium coil to chill the cold body and its contents to a temperature of 15 K.

The structural support for the suspension of the cryostat's cold body from the RT flange is attained by these tubes. The lengthening of access tubes through the bends might have decreased heat loss.

However, this would make the system more difficult to build since an extra structure would be needed to support the cold mass. Furthermore, due to the presence of other more contributing mechanisms, only a small reduction in heat loss can be realized. First, 1.6 liters of liquid nitrogen were pumped into the chilled body. Through LHe dewar, the flow of helium was controlled to keep the wall near 65 K above the freezing point of nitrogen (63.2 K) while the nitrogen was replenished from a nitrogen cylinder. This method allowed nitrogen to solidify into a fluid, resulting in a denser solid. This also keeps the container from being overfilled, which may cause detrimental tensions when the frozen object gets increasing temperature and on the expansion of SN2. For cooling up to 15 K, helium flow is increased. After the completion of tests, the temperature is allowed to get increase over

65°C, and up to the atmospheric pressure level, the pressure was increased. After the increase in the temperature of N_2 vapor up to room temperature, the volume of N_2 is quantified.

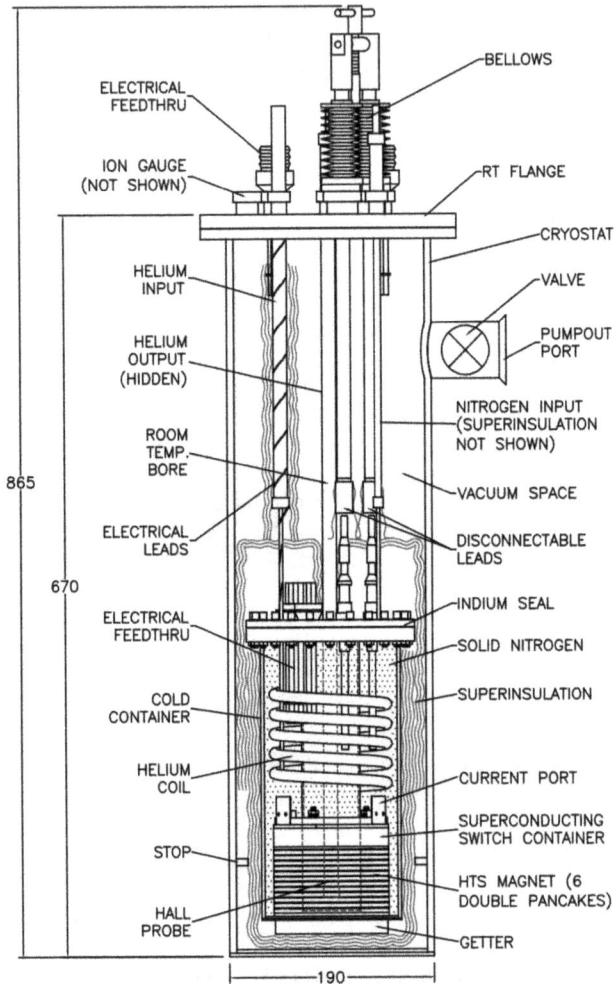

Fig. 1. Schematic drawing of the experimental apparatus used to operate the HTS magnet in thermal communication with solid nitrogen. The dimensions are given in m [7].

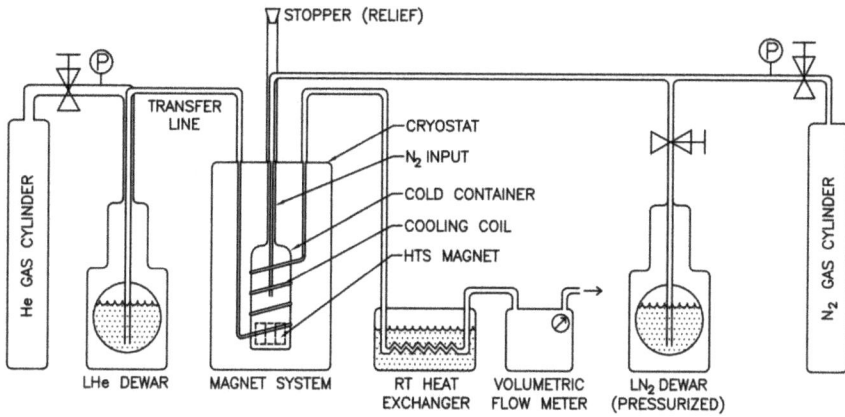

Figure 2. A method showing the cooling process to reduce temperature below 20 K [6, 7].

2.3 Magnesium diboride superconductors

An HTS is one feasible superconductor option for an LHe-free MRI magnet since its cooling source can function with cryocooler more easily than a NbTi magnet, which needs to function near 4.2-5.2 K [3]. It was considered that MgB_2, which had been found in January 2001 [8], would be an HTS that could compete with LTS in terms of cost and performance [9].

2.3.1 Challenges and prospects for MgB_2 MRI magnets

Although NbTi is a well-developed magnet and still the best magnet option for superconducting MRI systems, however, there are a couple of shortcomings that make this option a vulnerable one:

(1) Premature quenches produced by minute mechanical disturbances still occur, but considerably less frequently than in the 1980s; and

(2) Depend significantly on liquid helium.

These issues are being addressed to preserve NbTi MRI magnets as profitable products. Protection is a susceptible area for MgB_2 magnets, as it is for other HTS magnets [3].

There are two challenges the MgB_2 magnet has to overcome to break the NbTi barrier:

• MgB_2 wire, first. It needs to improve to the same level as NbTi in terms of magnetism. As a commercially accessible commodity, it must fulfill stringent magnet standards.

One of the magnet parameters is for a length of about 1 km, a difficulty for magnesium diboride wire. MgB_2 wire should not cost multiple times as much as NbTi wire.

Superconductors - Materials and Applications   Materials Research Forum LLC
Materials Research Foundations 132 (2022) 230-255   https://doi.org/10.21741/9781644902110-13

• Magnesium diboride MRI magnet is the second option. It won't be able to take the place of a NbTi MRI magnet. It must have enablers that eliminate the NbTi magnet's two flaws. These characteristics must be fully utilized. One apparent alternative is to use the MgB_2 MRI magnet, which is liquid helium free and operates at temperatures considerably over 4.2 K.

With roughly 35 000 MRI systems deployed globally and a rising industry anticipated reaching $7.4 billion per year [10, 11], medical diagnosis and therapy monitoring need the use of MRI. The most common full-body clinical MRI systems use 1.5 T or 3.0 T main magnetic field strengths and rely on liquid helium [12]. A liquid helium bath is used to immerse completely NbTi based MRI magnets that are typically operated at 4.2 K. A typical MRI system uses 1500-2000 l of LHe during its lifetime, with the primary cooling and testing taking place at the plant and subsequent magnet quenches taking place in the field [10]. Given the magnitude of the MRI business and its reliance on liquid helium, it's no surprise that the MRI sector accounts for 25% of global liquid helium demand [13, 14].

The price of LHe has risen significantly in recent years, and more price rises are likely this decade as demand grows faster than supply [11]. As LHe prices rise, the cost of building and maintaining LHe-cooled MRI equipment will grow as well. Furthermore, the scarcity of LHe in distant places and third-world nations makes it difficult for patients to obtain the equipment that radiologists prefer. Therefore, permanent magnet-based 0.2-0.5 T Low-field MRI systems are employed as liquid helium is not required for the operation.

Researchers have created and are testing an MRI that uses superconducting quantum interference devices (SQUID) and the magnetic field of Earth [15, 16]. As a result, the MRI community has been working hard to find and develop acceptable substitutes for liquid helium-cooled MRI primary magnets [14].

A 'dry' magnet that is cooled by conduction and has a two-stage cryocooler for refrigeration is an alternative to a 'wet' magnet. The cryocooler would only require 1-3 liters of LHe to function [16]. Nb_3Sn, MgB_2, YBCO, BSCCO, and Bi- 2223 having T_c higher than NbTi, 17 K, 39 K, 93 K, 108 K, 90-95 K respectively [17-20] are superconducting materials also feasible possibilities for magnet designs which are conduction-cooled. MgB_2 has the lowest T_c among the HTS alternatives for superconducting magnets. The wider operating temperature range of a higher T_c material improves the construction of the magnets which are conduction-cooled by enabling bigger temperature gradients as compared to that of NbTi. As a result, the reliance on liquid helium can be decreased, if not abolished entirely.

Being inexpensive, MgB_2 is a good candidate, also it has a somewhat high T_c, can be produced into persistent junctions, and is accessible in wire form [21-24]. The feasibility of MgB_2-based magnets has been an important subject of research concurrent with the development of MgB_2 wire [25-29]. For magnets with a field strength of 0.5 T, Monofilamentary MgB_2 wires were utilized to build coils [23]. MRI systems based on conduction-cooled NbTi have also been investigated [26].

Superconductors - Materials and Applications   Materials Research Forum LLC
Materials Research Foundations 132 (2022) 230-255   https://doi.org/10.21741/9781644902110-13

A 0.5 T, LHe-free MgB_2 magnet is used in commercial MR Open MRI. Because the magnet has an iron core and functions more in driven current, it is difficult to convert such a magnet design into one that is suited to MRI systems having 1.5 and 3.0 T magnets. The said magnets are functioned in a constant current manner (in which a permanent switch is required) without having iron contents to comply with MRI site weight limits.

Being operated in the current driven method, to ensure quench protection a dump resistor is used. A more advanced quench prevention mechanism that does not rely on the dump resistor is required for a sustained current mode magnet.

3. Potential superconductors for MRIs

Aside from NbTi and MgB_2, Nb_3Sn, BSCCO (2212, 2223), and REBCO have all been used to make superconducting wire in long enough lengths for magnet trials. Each of these superconductors might theoretically be used in 1.5 T and 3 T MRIs. They can work at greater temperatures than NbTi superconductors, allowing for complete conduction-cooled MRI magnets in the 4-10 K temperature range. Each, however, has drawbacks as compared to MgB_2 superconductors. The higher working temperatures of BSCCO (2212, 2223), and REBCO are offset by two challenges:

(1) MRI persistent coils and joints necessitate the fabrication of persistent joints at a high degree of quality.

(2) Price performance in the range of 4-10 K and 4-6 T.

In comparison to NbTi and anticipated MgB_2 wires, the cost for the three mentioned materials is larger and seems to be in the future. Nb_3Sn wires have been shown to produce MRI-quality persistent joints, and great progress has been made in enhancing the J_c of Nb_3Sn wires during the previous ten years. For the International Fusion Program (ITER), large quantities of Nb_3Sn superconductor wires have been produced [30].

3.1 Nb-Ti and Nb_3Sn superconductors

NbTi wires were completely industrialized in 1968, just a few years after their discovery, and are now widely employed in practical applications. Since the materials and fabrication costs are significantly lesser, NbTi alloy remains the most cost-effective superconducting material for applications based on liquid helium. Nb_3Sn, on the other hand, is not feasible to get into different shapes like Nb-Ti alloy due to the brittleness of the ingredients.

A superconducting phase of Nb_3Sn can be created by heat treatment through the reaction of Sn and Nb. Increased pinning capability is the key to increasing Nb_3Sn wires' non-Cu J_c [31]. Because of the additional complexity of wire fabrication, Nb_3Sn superconducting wires were first commercialized after 1970, and are more expensive than Nb-Ti superconductors. 9 T and 11 T fields at 4.2 K and 1.8 K respectively can be generated through superconducting magnets made of Nb-Ti. Several applications consume thousands of tonnes of Nb-Ti superconducting wires each year. The market niche for Nb_3Sn superconductors is high-field applications in the range of up to 23 T, which are beyond the capabilities of Nb-Ti. For example, from 2008 to 2015, around 500 tonnes of Nb_3Sn wires were acquired for the ITER project, resulting in several times rise in Nb_3Sn manufacturing capabilities globally [32].

3.2   Copper based superconductors

A good aspect of cuprate superconductors is their higher T_c, even higher than liquid nitrogen (77 K). Since nitrogen is abundant, therefore, cooling with liquid nitrogen is far less expensive than cooling with liquid helium. Three compounds are having higher T_c including Bi-2223, Bi-2212, and REBCO, yet their ceramic brittleness makes the manufacturing of wires and tapes far more challenging. After high-temperature heat treatment, cuprate compounds form plate-like crystals, however, to achieve the best superconducting performance, oxygen content must be carefully controlled. Furthermore, with wide grain boundary angles, a weak-joining effect among the grains exists. This is not good for the capacity to carry the and can be decreased by grain texture. For large inter-grain critical currents in Bi-2223 superconductors, well-oriented grains with a c-axis misorientation of 15° are required. PIT wires or tapes for Bi-2212 superconductors can be created because well-oriented and aligned grains are achievable from the molten phase at lower temperatures. The PIT method is incompatible with REBCO because it can only carry significant critical current [33-36]. Silver or its alloys are required as cover for BSCCO instead of Cu and other metals because they are chemically well-suited. Multifilamentary wires can be achieved however it is necessary to align the arbitrarily oriented BSCCO granules to increase the connection between them. Because of their great anisotropy, BSCCO oxides crystallize with a plate-like appearance, and the orientation of grain is very simple. Bi-2212 and 2223 on the other hand, use distinct methods for grain orientation. Partially melt and then progressive cooling is used for Bi-2212 wires, whereas rolling induces texturing is used for Bi-2223 tapes.

Sumitomo [37] developed a controlled over-pressure technique in 2005 that reduced porosity and enhanced grain connectivity, resulting in an increase in production yield and an I_c of 200 A at 77 K, self-field. While the transport current density (J_c) for Bi-2212 wires was not changed considerably until 2011, later it was discovered that porous regions in the ceramic, instead of grain misorientation, were the principal barrier to supercurrent transport [38]. High-pressure heating can lead to twice increased transport J_c of Bi-2212 wires up to 4105 A/cm^2 [39]. Furthermore, a homogenous precursor powder with minimal impurities can aid in the reduction of porosity, increasing the transport J_c to 6.6105 A/cm^2 [40,41].

Sumitomo's laminated mechanical reinforcing approach greatly improves the mechanical strength of the weak Ag encased Bi-2223 tapes, making them a viable option for high-field applications [42,43]. However, in recent years, REBCO has attracted a growing amount of attention, and the study on Bi-2223 has dwindled [44].

3.3   Rare – earth barium copper oxide superconductors (REBCO)

At 77 K, REBCO possesses a substantially lesser anisotropy and greater in-field J_c than BSCCO superconductors. Nevertheless, for REBCO the uniaxially lined-up grains created in BSCCO wires are insufficient to achieve large J_c, hence making it important to have biaxial grains. A few meters thick silver and heavier copper layers are also deposited on the conductor for environmental protection and thermal stability. Since 2003, a variety of methods for generating coated conductors have been used in industry [45-48]. Both chemical procedures, as well as physical methods, can be used to build epitaxial REBCO layers (RCE). Ex-situ procedures such as MOD and RCE, for example, involve the deposition of pre-cursors followed by the transformation of the precursors to REBCO. Physical methods support both deposition and formation simultaneously. PLD is costlier and has a slower deposition rate. Despite their rapid deposition rates and large deposition areas, MOCVD and RCE products exhibit poor pinning microstructures and crystallinity. Depending on their performance, coated conductors made via various approaches can be employed for a variety of applications [49-51].

REBCO-coated conductors with good performance and longer lengths have been commercialized in recent years by companies in the United States, Japan, Korea, Germany, Russia, and China. At 77 K, their commercial coated conductors have I_c values ranging from 100 to over 250 A [48,52,53]. Artificial pinning centers can change the J_c anisotropy, and they can further decrease the anisotropy of REBCO tapes [48,54,55]. It was also discovered that by increasing the thickness of superconducting layers, APCs can prevent J_c deterioration in REBCO tapes [51,56,57]. Aside from power application projects, high-field applications utilizing coated conductors are progressing steadily. DC magnetic field of 45.5 T with 14.4 T REBCO insertion in a 31.1 T magnet has been achieved in the past [58]. RIKEN [59], NHMFL [60], and IEECAS [61] also developed the same type of products with improving features and performances. These endeavors to achieve better performance with stronger magnetic fields demonstrate REBCO tapes' enormous potential in the industry.

3.4 MgB₂ superconductors

The superconducting critical transition temperature of MgB_2 was 39K in 2001, which set a new record for T_c. Unlike copper oxide high-temperature superconductors, MgB_2 superconductor currents are not affected by weakly connected grain boundaries [62], making MgB_2 a viable material for constructing high-performance wires. On the other hand, due to its poor flux pinning ability, MgB_2's critical current density reduces fast as the applied magnetic field increases, limiting its use in high magnetic field regions. MgB2 superconductor has gotten a lot of attention from the applied superconductivity field for the reason of having comparatively high T_c, inexpensive cost of material, lightweight, and easy chemical conformation. Among the superconducting magnets used in MRI systems, it is widely assumed that MgB_2 superconducting materials offer apparent technological and financial benefits [63].

The rising field limited the current carrying capacity of MgB_2 superconductors owing to a shortage of pinning centers. Substitution of Carbon is now the utmost effective and extensively utilized approaches for increasing J_c. The substituted carbon in the MgB_2 lattice can take the position of the boron site, lowering the mean free path of electrons and raising the upper critical field. Nano-SiC and carbon-containing compounds are the most effective carbon dopants [63-65].

With carbon addition, J_c of MgB_2 wires developed through the PIT technique is possibly enhanced to 6104 A/cm^2 at 4.2 K and 10 T [66]. Several labs have attained 100 m MgB_2 wires using the IMD method [67], while several industries have produced kilometer-long practical MgB_2 wires using the PIT method.

Fig. 3 shows typical critical current and current density measurements vs. applied field at various temperatures for the standard multifilament MgB_2 conductor manufactured at Hyper Tech. Typical J_c values are 1×10^5 A cm^{-2} at 5 T, 4.2 K, and 2×10^5 A cm^{-2} at 1 T, 20 K. The I_c at 1 T and 20 K is approximately 200 A [68].

Hyper Tech and Columbus' MgB_2 wires were used in many applications including Fault current limiters, Magnetic resonance imaging, and Superconducting cables. The first MgB_2 open MRI has been developed by Columbus in 2006 [69]. They developed more than 20 sets of equipment mentioned before. The operating field strength is expected to grow to 1-2 T in the future.

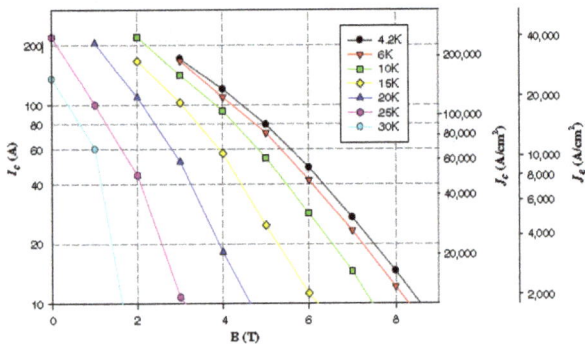

Fig. 3. Typical I_c, J_e, and J_c properties for the standard Hyper Tech multifilament MgB_2 strand. Data on graph is limited to $J_c > 104$ A cm^{-2}.

3.5   Iron-based superconductors (IBS)

In IBS, currents among the misaligned grains depreciate to a smaller degree as compared to REBCO [70]. As a consequence, the inexpensive PIT technique, utilized to produce marketable Nb_3Sn, Bi-2223 and MgB_2 wires, appears to have promise for IBS wire manufacturing. In contrast to BSCCO wires, which were restricted to silver or its alloys, IBS wires had a wider range of sheath materials to choose from. For IBS wires, for example, a composite sheath made of various inexpensive and sturdy metals for the outside cover and Ag as the inside one may be utilized. The composite sheath may lower the Ag cost ratio, and improve chemical stability and mechanical qualities.

Superconductors - Materials and Applications
Materials Research Foundations 132 (2022) 230-255

Materials Research Forum LLC
https://doi.org/10.21741/9781644902110-13

In 2008, the IEECAS used the in situ PIT technique to make the first IBS wires. The ex-situ PIT technique considerably increases the density and phase uniformity of iron pnictides wires following final heat treatment [71]. Recently, mechanical deformation processes in the United States, Europe, China, Japan, and Australia have enhanced the J_c of 122-type IBS [72-78].

Better grain alignment and texture have been achieved through the hot press method, resulting in enhanced J_c in Ba-122 tapes up to 1.5105 A/cm^2 (I_c = 437 A), displaying extremely minor field dependency up to 33 T [79]. At 4.2 K and 10 T, a high J_c of 4104 A/cm^2 was obtained for Cu/Ag sheathed Ba-122 round wire manufactured utilizing hot isostatic pressing densification (HIP) [80]. The groove rolling process gives a local grain alignment perpendicular to the wire axis [81,82]. Table 1 shows the comparison of important properties according to their properties of three high T_c superconductors.

Table 1 Comparison of three representative high-T_c superconductors.

	IBSCs	MgB$_2$	Cuprates
Parent material	AFM semimetal (T_N ~ 150 K)	Pauli paramagnetic metal	AFM Mott insulator (T_N ~ 400 K)
Fermi level	Fe 3d 5-orbitals	B 2p 2-orbitals	Cu 3d single orbital
Maximum T_c (K)	55 (for 1111 type), 38 (for 122 type)	39	93 (YBCO), 110 (Bi2223)
Impurity	Robust	Sensitive	Sensitive
SC gap symmetry	Extended s-wave	s-wave	d-wave
Upper critical field at 0 K, $H_{c2}(0)$ (T)	100-200 (for 1111 type) 50-100 (for 122 type) ~ 50 (for 11 type)	40	> 100
Crystallographic symmetry in SC state	Tetragonal	Hexagonal	Orthorhombic (Y-and Bi-systems)
Anisotropy, γ	4-5 (for 1111 type) 1–2 (for 122-and 11-types)	2	5-7 (YBCO), 50-90 (Bi-system)
Irreversibility field, H_{irr} (T)	> 50 (4 K) > 15 (20 K)	> 25 (4 K) > 10 (20 K)	> 0 (77 K, YBCO)
Critical GB angle, θ_c (deg.)	8-9	No data	3-5 (YBCO)
Advantage	High $H_{c2}(0)$, Easy fabrication	Easy fabrication	High T_c and $H_{c2}(0)$
Disadvantage	Toxicity	Low $H_{c2}(0)$	High cost due to 3D alignment of crystallites

Since Fe has proven to be chemically well-suited with the 11-IBS, it is the most often used sheath material for 11-type Fe chalcogenides wires produced using the conventional PIT approach [77]. The thin-film approach, on the other hand, shows how it may be used to make 11-IBS coated conductors. Because of the simple elemental composition, moderate T_c (16 K), high upper critical fields, and relative ease of manufacture for 11-IBS thin films, it is particularly promising for high-field applications at low temperatures [83]. Additionally, the newly discovered iron chalcogenides $K_xFe_2Se_2$ ($T_c > 30$ K) and FeSe-based (Li, Fe) OHFeSe ($T_c = 40$ K) probably widen their uses [84,85].

IEECAS [86] successfully generated Fe/Ag clad 7-filament 122-IBS wires and tapes using the PIT approach in 2013 [87]. After that, Fe/Ag encased 114-filament 122-IBS wires and tapes have been developed [88]. 122-IBS tape was created by IEECAS team employing a scalable rolling technique to achieve the length of 11 m in 2014. Subsequently, developing the longer wire production procedure to realize a larger degree of deformation uniformity, the same group produced IBS tapes having a length of 100 m [89].

In 2020, the IBS racetrack coil was first manufactured by the Chinese Academy of Sciences' Institute of High Energy Physics. The best IBS racetrack coils had an operational current of 65 A when quenched at 4.2 K. This is still greater than 86.7 percent of smaller samples at 10 T [90]. Figure 4 depicts these findings, manifesting the potential of IBS conductors for high strength field magnet applications, notably in future high-energy accelerators [77, 90, 91].

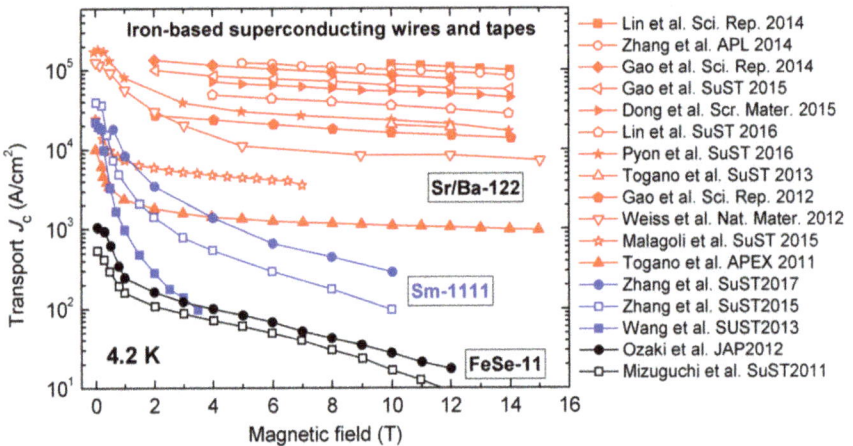

Fig. 4 At 4.2 K, variation in J_c in PIT based 122, 1111 and 11 wires and tapes against magnetic field.

Four major types of IBSCs are:

1. SmOFeAs (T_c = 55 K, 1111 type)

2. Sr/BaFe$_2$As$_2$ (T_c = 38 K, 122 type)

3. LiFeAs (T_c = 18 K, 111 type)

4. FeSe (T_c = 8 K, 11 type)

From an application point of view, the 122 type is the best-optimized one based on its ultra-high H_{c2} > 70 T at a temperature of 20 K, additionally having lower anisotropy values (γ < 2) and convenient route of fabrication.

Researchers from IEECAS were the initiators to report 1111 and 122 prototype wires through the PIT process just after the discovery of superconductivity in pnictides [92,93]. It was clarified afterward by the Hosono group that in 122 pnictides the weak-link effect was not as heavy as it is in YBCO [94]. This leads to the realization that the scalable PIT method is applicable for the preparation of pnictides wires and tapes.

Higher J_c up to 6.1×10^4 and 3.5×10^4 A/cm^2 at 4.2 K and 10 T have been found in Sr-122 tapes with 7 and 19 core respectively [96]. As shown in fig. 5, the same research team developed 114-filament Sr-122/Ag/ Fe conductors attaining less than 50 μm cross-sectional filament size by converting the conductors into wires having 2.0 mm diameters. Reduced tape thickness causes the increase in transport J_c owing to the conversion of round wires into flat tapes. Transport J_c depends on the field to a lesser extent for these 114-filament samples up to 14 T, no matter they possess different shapes and dimensions. This observation indicates their possible implementation in stronger field applications.

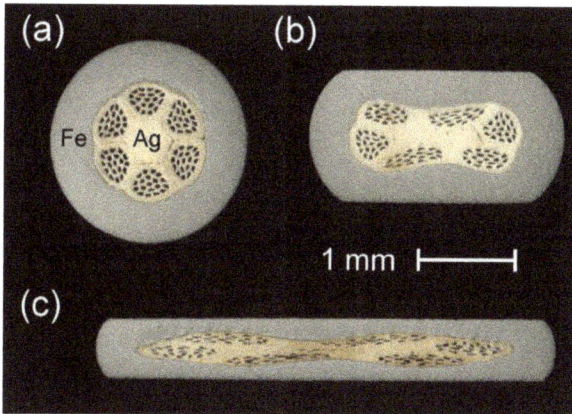

Fig. 5 Employing scalable PIT route, cross sectional view of multi-filamentary 122-type IBSC [95].

The T_c magnetic field relation's feasible application space is highlighted in fig. 6. High magnetic fields are one of the most promising applications for superconducting wires. The nature of IBSCs makes them resistant to greater magnetic fields. The magnetic field dependency of J_c of IBSC has recently been practically verified, demonstrating its outstanding tolerance to higher magnetic fields. Superconductors may now operate at 20-30 K without the need for liquid He, thanks to significant improvements in refrigeration system performance. This scenario offers the possibility of using current IBSCs in greater magnetic fields. A small NMR device based on a superconducting magnet independent of Liquid-He will likewise be a useful milestone [76].

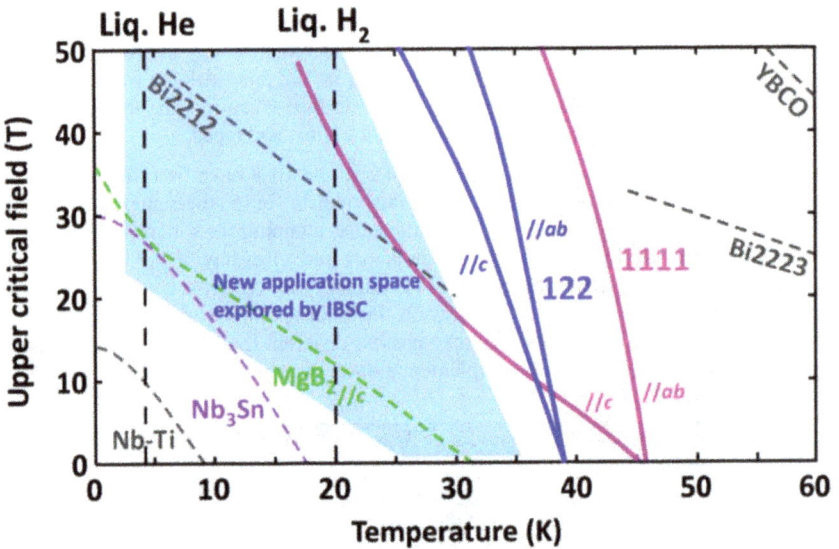

Fig: 6 A comparative analysis of higher critical H_{c2} for various high field superconductors [98] including $NdFeAsO_{1-x}F_x$ [99], $Ba_{1-x}K_xFe_2As_2$ (122) [100], YBCO [101], Bi2212 [101] and Bi2223 [102].

4. Materials' and their applications' prospects in the future

Commercial demands of magnets with high field strength mainly including MRI and NMR are currently driving the development of higher-performance superconducting materials. The capacity to produce magnetic fields that are not limited by the constraints of LTS conductors expands the possibilities for HTS materials and speeds up their industrial production. Cost, technical maturity, and batch stability are advantages of LTS materials over HTS materials.

MRI and NMR systems are being developed drastically in commercial applications using superconducting magnets, which has resulted in a comparable exponential expansion in wire manufacturing capacity. A greater than 14 T whole-body MRI will be the next target incorporating Nb_3Sn superconducting wires instead of Nb-Ti [103].

The cost of superconductors becomes a critical concern for the above-mentioned systems and other massive projects too. The product price remains the key barrier to large-scale application for copper oxide superconductors which are close to being commercialized. Nb_3Sn conductors cost roughly 5 $/kA m, Bi-2212 & 2223 conductors cost 60-80 $/kA m, and REBCO conductors cost 100-200 $/kA m. For large-scale applications, their cost is still above the desired 25 $/kA m. As a result, their research and development efforts are focused on decreasing the price. The low manufacturing yield of long-length tape goods, particularly those longer than 500 m, appears to be one of the reasons behind REBCO coated conductors' high cost [35,104].

Aside from the expense, the comparatively low mechanical strength of wires and tapes comprising Bi-2223 and Bi-2212 owing to the existence of soft silver or silver alloys sheaths is another key factor to consider for applications with strong electromagnetic stress. Heat treatment, on the other hand, can lower the mechanical strength of the magnet, which has become a serious hurdle for up to 40 T magnets. Sumitomo has successfully used the laminated mechanical reinforcing process for commercial Bi-2223 tapes [105,106].

Conclusion

Magnetic resonance imaging and nuclear resonance imaging remain the most profitable applications of superconducting materials to date. All the superconducting materials that have been developed for the applications of MRI have strengths and weaknesses as some of these materials are LTS and some are HTS. Some of them require cooling systems for their better performances and some require different working conditions. Researchers have been working to reduce the cost of MRI magnets as LTS require liquid Helium which increases the cost of MRI. Similarly, there are HTS that can be operated without liquid helium or using the conduction cooling that can reduce the cost of MRI magnets for future applications. Wires and joints of different MRI magnets such as MgB_2 and Nb_3Sn has been an important subject of research for commercial applications of MRI magnets.

References

[1] P. A. Abetti and P. Haldar, One hundred years of superconductivity: science, technology, products, profits and industry structure, Int. J. Technol. Management, 48 (2009) 423-447. https://doi.org/10.1504/IJTM.2009.026688

[2] T. C. Cosmus and M. Parizh, Advances in whole-body MRI magnets, IEEE Trans. Appl. Supercond. 21 (2011) 2104 - 2109. https://doi.org/10.1109/TASC.2010.2084981

[3] Y. Iwasa, Case Studies in Superconducting Magnet 2nd ed., Springer, Berlin, 2009. https://doi.org/10.1007/b112047_1

Superconductors - Materials and Applications   Materials Research Forum LLC
Materials Research Foundations 132 (2022) 230-255   https://doi.org/10.21741/9781644902110-13

[4] Y. Iwasa, Design and operational issues for 77 K superconducting magnets, IEEE Trans. Mag. 24 (1988) 1211. https://doi.org/10.1109/20.11452

[5] Y. Iwasa, A 'permanent' HTS magnet system: key design and operational issues, Advances in Superconductivity, Springer, Tokyo, 1998. https://doi.org/10.1007/978-4-431-66879-4_325

[6] B. J. Haid, A 'permanent' high-temperature superconducting magnet operated in thermal communication with a mass of solid nitrogen, Ph.D. Thesis, Department of Mechanical Engineering, MIT, Cambridge, 2001.

[7] B. J. Haid, H. Lee, Y. Iwasa, S. S. Oh, Y. K. Kwon, K. S. Ryu, A 'permanent' high-temperature superconducting magnet operated in thermal communication with a mass of solid nitrogen, Cryogenics 42 (2002) 229-244. https://doi.org/10.1016/S0011-2275(02)00022-X

[8] J. Nagamatsu, N. Nakagawa, T. Muranaka, Y. Zenitani and J. Akimitsu, Superconductivity at 39 K in magnesium diboride, Nature, 410 (2001) 63-64. https://doi.org/10.1038/35065039

[9] D. Larbalestier, A. Gurevich, D. M. Feldmann, and A. Polyanskii, High-Tc superconducting materials for electric power applications. In Materials For Sustainable Energy: A Collection of Peer-Reviewed Research and Review Articles from Nature Publishing Group (2011) 311-320. https://doi.org/10.1142/9789814317665_0046

[10] Y. Lvovsky, E. W. Stautner, and T. Zhang, Novel technologies and configurations of superconducting magnets for MRI, Supercond. Sci. Technol. 26 (2013) 93001. https://doi.org/10.1088/0953-2048/26/9/093001

[11] Y. Lvovsky and P. Jarvis, Superconducting systems for MRI-present solutions and new trends, IEEE Trans. Appl. Supercond. 15 (2005) 1317-1325. https://doi.org/10.1109/TASC.2005.849580

[12] C. Spencer, P. A. Sanger and M. Young, The temperature and magnetic field dependence of superconducting critical current densities of multifilamentary Nb3Sn and NbTi composite wires, IEEE Trans. Magn. 15 (1979) 76-79. https://doi.org/10.1109/TMAG.1979.1060146

[13] W. Nuttall, R. Clarke and B. Glowacki, The Future of Helium as a Natural Resource 1st ed, Routledge, Abingdon, UK, 2014.

[14] Z. Cai, R. H. Clarke, B. A. Glowacki, W. J. Nuttall, and N. Ward, Ongoing ascent to the helium production plateau- insights from system dynamics, Resour. Policy, 35 (2010) 77-89. https://doi.org/10.1016/j.resourpol.2009.10.002

[15] M. A. Espy, P. E. Magnelind, A. N. Matlashov, S. G. Newman, H. J. Sandin, L. J. Schultz, R. Sedillo, A. V. Urbaitis and P. L. Volegov, Progress toward a deployable SQUID based ultra-low field MRI system for anatomical imaging, IEEE Trans. Appl. Supercond. 25 (2015) 1-5. https://doi.org/10.1109/TASC.2014.2365473

[16] H. Morita, M. Okada, K. Tanaka, J. Sato, H. Kitaguchi, H. Kumakura, K. Togano, K. Itoh, and H. Wada, 10 T conduction-cooled Bi-2212/Ag HTS solenoid magnet system, IEEE Trans. Appl. Supercond. 11 (2001) 2523-2526. https://doi.org/10.1109/77.920379

[17] T. Baig, Z. Yao, D. Doll, M. Tomsic, and M. Martens, Conduction cooled magnet design for 1.5 T, 3.0 T, and 7.0 T MRI systems, Supercond. Sci. Technol. 27 (2014) 125012. https://doi.org/10.1088/0953-2048/27/12/125012

[18] H. Miyazaki, S. Iwai, T. Tosaka, K. Tasaki, S. Hanai, M. Urata, S. Ioka, and Y. Ishii, Development of a 5.1 T conduction-cooled YBCO coil composed of a stack of 12 single Pancakes, Physica C, 484 (2013) 287-291. https://doi.org/10.1016/j.physc.2012.02.041

[19] M. A. Young, J. A. Demko, M. J. Gouge, M. O. Pace, J. W. Lue, and R. Grabovickic, Measurements of the performance of BSCCO HTS tape under magnetic fields with a cryocooled test rig, IEEE Trans. Appl. Supercond. 13 (2003) 2964-2967. https://doi.org/10.1109/TASC.2003.812076

[20] Q. Wang, S. Song, Y. Lei, Y. Dai, B. Zhang, C. Wang, S. Lee and K. Kim, Design and fabrication of a conduction-cooled high-temperature superconducting magnet for 10 kJ superconducting magnetic energy storage system, IEEE Trans. Appl. Supercond. 16 (2006) 570-573. https://doi.org/10.1109/TASC.2005.869683

[21] S. L. Budko, G. Lapertot, C. Petrovic, C. E. Cunningham, N. Anderson, and P. C. Canfield, Boron isotope effect in superconducting MgB2 Phys. Rev. Lett. 86 (2001) 1877-1880. https://doi.org/10.1103/PhysRevLett.86.1877

[22] J. Nagamatsu, N. Nakagawa, T. Muranaka, Y. Zenitani and J. Akimitsu, Superconductivity at 39 K in magnesium diboride, Nature 410 (2001) 63-64. https://doi.org/10.1038/35065039

[23] J. Ling, J. Voccio, Y. Kim, S. Hahn, J. Bascunan, D. K. Park and Y. Iwasa, Monofilament wire for a whole-body MRI magnet: superconducting joints and test coils, IEEE Trans. Appl. Supercond. 23 (2013) 6200304. https://doi.org/10.1109/TASC.2012.2234183

[24] M. Tomsic, M. Rindfleisch, J. Yue, K. McFadden, J. Phillips, M. D. Sumption, M. Bhatia, S. Bohnenstiehl and E. W. Collings, Overview of MgB2 superconductor applications, Int. J. Appl. Ceram. Technol. 4 (2007) 250-259. https://doi.org/10.1111/j.1744-7402.2007.02138.x

[25] S. Mine, M. Xu, S. Buresh, W. Stautner, C. Immer, E. T. Laskaris, K. Amm, and G. Grasso, Second test coil for the development of a compact 3 T magnet, IEEE Trans. Appl. Supercond. 23 (2013) 4601404. https://doi.org/10.1109/TASC.2012.2232692

[26] Yao, Weijun, J. Bascunan, W-S. Kim, S. Hahn, H. Lee, and Y. Iwasa, A solid nitrogen-cooled demonstration: coil for MRI applications, IEEE Trans. Appl. Supercond. 18 (2008) 912-915. https://doi.org/10.1109/TASC.2008.920836

Superconductors - Materials and Applications   Materials Research Forum LLC
Materials Research Foundations 132 (2022) 230-255   https://doi.org/10.21741/9781644902110-13

[27] J. Bascunan, H. Lee, E. S. Bobrov, S. Hahn, Y. Iwasa, M. Tomsic and M. Rindfleisch, A 0.6 T/650 mm RT bore solid nitrogen cooled demonstration coil for MRI a status report, IEEE Trans. Appl. Supercond. 16 (2006) 1427-1430. https://doi.org/10.1109/TASC.2005.864456

[28] W. Yao, J. Bascuan, S. Hahn and Y. Iwasa, MgB2 Coils for MRI Applications, IEEE Trans. Appl. Supercond. 20 (2010) 756-759. https://doi.org/10.1109/TASC.2009.2038890

[29] D. K. Park, S. Hahn, J. Bascunan and Y. Iwasa, Active protection of an MgB2 test coil, IEEE Trans. Appl. Supercond. 21 (2011) 2402-2405. https://doi.org/10.1109/TASC.2010.2095812

[30] D. Bessette and N. Mitchell, Review of the results of the ITER toroidal field conductor R & D and qualification, IEEE Trans. Appl. Supercond. 18 (2008) 1109-1113. https://doi.org/10.1109/TASC.2008.921277

[31] X. Xu, M. Sumption, X. Peng, and E. W. Collings, Refinement of Nb3Sn grain size by the generation of ZrO2 precipitates in Nb3Sn wires, Appl. Phys. Lett. 104 (2014) 82602. https://doi.org/10.1063/1.4866865

[32] X. Xu, A review and prospects for Nb3Sn superconductor development, Supercond. Sci. Technol. 30 (2017) 093001. https://doi.org/10.1088/1361-6668/aa7976

[33] N. Banno Y. Miyamoto, K. Tachikawa, Multifilamentary Nb3Sn Wires Fabricated Through Internal Diffusion Process Using Brass Matrix, IEEE Trans. Appl. Supercond. 26 (2016) 6001504. https://doi.org/10.1109/TASC.2016.2531123

[34] X. Xu, M. D. Sumption, and X. Peng, Internally oxidized Nb3Sn superconductor with very fine grain size and high critical current density, Adv. Mater. 27 (2015) 1346. https://doi.org/10.1002/adma.201404335

[35] S. Balachandran, C. Tarantini, P. J. Lee, F. Kametani, Y. F. Su, B. Walkker, W. L. Starch and D. B. Larbalestier, Beneficial influence of Hf and Zr additions to Nb4at%Ta on the vortex pinning of Nb3Sn with and without an O source, Supercond. Sci. Technol. 32 (2019) 044006. https://doi.org/10.1088/1361-6668/aaff02

[36] D. Uglietti, A review of commercial high-temperature superconducting materials for large magnets: from wires and tapes to cables and conductors, Supercond. Sci. Technol. 32 (2019) 053001. https://doi.org/10.1088/1361-6668/ab06a2

[37] S. Kobayashi, K. Yamazaki, T. Kato, K. Ohkura, E. Ueno, K. Fujino, J. Fujikami, N. Ayai, M. Kikuchi, K. Hayashi and K. Sao, Controlled over-pressure sintering process of Bi2223 wires, Physica C: Superconductivity and its applications, 426 (2005) 1132-1137. https://doi.org/10.1016/j.physc.2005.02.097

[38] F. Kametani, T. Shen, J. Jiang, C. Scheuerlein, A. Malagoli, M. Di Michiel, Y. Huang, H. Miao, J. A. Parrell, E. E. Hellstrom and D. C. Larbalestier, Bubble formation within filaments of melt-processed Bi2212 wires and its strongly negative

effect on the critical current density, Supercond. Sci. Technol. 24 (2011) 075009. https://doi.org/10.1088/0953-2048/24/7/075009

[39] J. Jiang, W. L. Starch, M. Hannion, F. Kametani, U. P. Trociewitz, E. E. Hellstrom and D. C. Larbalestier, Doubled critical current density in Bi-2212 round wires by reduction of the residual bubble density, Supercond. Sci. Technol. 24 (2011) 082001. https://doi.org/10.1088/0953-2048/24/8/082001

[40] J. Jiang, G. Bradford, S. I. Hossain, M. D. Brown, J. Cooper, E. Miller, Y. Huang, H. Miao, J. A. Parrell, M. White, and A. Hunt, High-performance Bi-2212 round wires made with recent powders, IEEE Trans. Appl. Supercond. 29 (2019) 1-5. https://doi.org/10.1109/TASC.2019.2895197

[41] K. Sato, S. Kobayashi, and T. Nakashima, Present status and future perspective of bismuth-based high-temperature superconducting wires realizing application systems, Jpn. J. Appl. Phys. 51 (2011) 010006. https://doi.org/10.1143/JJAP.51.010006

[42] L. Xiao, S. Dai, L. Lin, J. Zhang, W. Guo, D. Zhang, Z. Gao, N. Song, Y. Teng, Z. Zhu, Z. Zhang, G. Zhang, F. Zhang, X. Xu, W. Zhou, Q. Qiu, and H. Li, Development of the World's First HTS Power Substation, IEEE Trans. Appl. Supercond. 22 (2012) 5000104 https://doi.org/10.1109/TASC.2011.2176089

[43] S. Awaji, K. Watanabe, H. Oguro, H. Miyazaki, S. Hanai, T. Tosaka, and S. Loka, First performance test of a 25 T cryogen-free superconducting magnet, Supercond. Sci. Technol. 30 (2017) 065001. https://doi.org/10.1088/1361-6668/aa6676

[44] T. Nakashima, S. Kobayashi, T. Kagiyama, M. Yamazaki, S. Kikuchi, K. Yamade, K. Hayashi, K. Sato, G. Osabe, and J. Fuikami, Overview of the recent performance of DI-BSCCO wire, Cryogenics 52 (2012) 713-718. https://doi.org/10.1016/j.cryogenics.2012.06.018

[45] D. C. Larbalestier, J. Jiang, U. A. Trociewitz, F. Kametani, C. Scheuerlein, M. D. Canassy, M. Matras, P. Chen, N. C. Craig, P. J. Lee, and E. E. Hellstrom, Isotropic round-wire multifilament cuprate superconductor for generation of magnetic fields above 30 T, Nat. Mater. 13 (2014) 375-381. https://doi.org/10.1038/nmat3887

[46] J. Zheng, Y. Song, X. Liu, J. Li, Y. Wan, M. Ye, K. Ding, S. Wu, W. Xu and J. Wei, Concept design of hybrid superconducting magnet for CFETR tokamak reactor, Proc. IEEE 25th SOFE (2013) 1-6. https://doi.org/10.1109/SOFE.2013.6635364

[47] J. G. Qin, Y. Wu, J. G. Li, C. Dai, F. Liu, H. J. Liu, P. H. Liu, C. S. Li, Q. B. Hao, C. Zhou and S. Liu, Manufacture and test of Bi-2212 cable-in-conduit conductor, IEEE Trans. Appl. Supercond. 27 (2017) 1-5. https://doi.org/10.1109/TASC.2017.2652306

[48] C. Senatore, M. Alessandrini, A. Lucarelli, R. Tediosi, D. Uglietti, and Y. Iwasa, Progresses and challenges in the development of high-field solenoidal magnets based on RE123 coated conductors, Supercond. Sci. Technol. 27 (2014) 103001. https://doi.org/10.1088/0953-2048/27/10/103001

Superconductors - Materials and Applications   Materials Research Forum LLC
Materials Research Foundations 132 (2022) 230-255   https://doi.org/10.21741/9781644902110-13

[49] X. Obradors and T. Puig, Coated conductors for power applications: materials challenges, Supercond. Sci. Technol. 27 (2014) 044003. https://doi.org/10.1088/0953-2048/27/4/044003

[50] J. L. MacManus-Driscoll and S. C. Wimbush, Processing and application of high-temperature superconducting coated conductors, Nat. Rev. Mater. 6 (2021) 587-604. https://doi.org/10.1038/s41578-021-00290-3

[51] Y. Shiohara, T. Taneda, and M. Yoshizumi, Overview of Materials and Power Applications of Coated Conductors Project, Jpn. J. Appl. Phys. 51 (2012) 010007. https://doi.org/10.7567/JJAP.51.010007

[52] C. Senatore, C. Barth, M. Bonura, M. Kulich and G. Mondonico, Field and temperature scaling of the critical current density in commercial REBCO coated conductors, Supercond. Sci. Technol. 29 (2016) 014002. https://doi.org/10.1088/0953-2048/29/1/014002

[53] M. Durrschnabel, Z. Aabdin, M. Bauer, R. Semerad, W. Prusseit, and O. Eibl, DyBa2Cu3O7−x superconducting coated conductors with critical currents exceeding 1000 A cm−1, Supercond. Sci. Technol. 25 (2012) 105007. https://doi.org/10.1088/0953-2048/25/10/105007

[54] H. S. Kim, S. S. Oh, H. S. Ha, D. Youm, S. H. Moon, J. H. Kim, S. X. Dou, Y. U. Heo, S. H. Wee, and A. Goyal, Ultra-high performance, high-temperature superconducting wires via cost-effective, scalable, co-evaporation process, Scientific reports, Sci. 4 (2014) 1-6. https://doi.org/10.1038/srep04744

[55] G. Majkic, R. Pratap, A. Xu, E. Galstyan, H. C. Higley, S. O. Prestemon, X. Wang, D. Abraimov, J. Jaroszynski, and V. Selvamanickam, Engineering current density over 5 Ka mm−2 at 4.2 K, 14 T in thick film REBCO tapes, Supercond. Sci. Technol. 31 (2018)10LT01. https://doi.org/10.1088/1361-6668/aad844

[56] K. Zhang, S. Hellmann, M. Calvi, T. Schmidt, Magnetization Simulation of Rebco Tape Stack With a Large Number of Layers Using the Ansys A-V-A Formulation, IEEE Trans. Appl. Supercond. 30 (2020) 4700805. https://doi.org/10.1109/TASC.2020.2965506

[57] C. Lee, H. Son, Y. Won, Y. Kim, C. Ryu, M. Park, and M. Iwakuma, Progress of the first commercial project of high-temperature superconducting cables by KEPCO in Korea, Supercond. Sci. Technol. 33 (2020) 044006. https://doi.org/10.1088/1361-6668/ab6ec3

[58] S. Hahn, K. Kim, K. Kim, X. Hu, T. Painter, I. Dixon, S. Kim, K. R. Bhattarai, S. Noguchi, J. Jaroszynski, and D. C. Larbalestier, 45.5-tesla direct-current magnetic field generated with a high-temperature superconducting magnet, Nature 570 (2019) 496-499. https://doi.org/10.1038/s41586-019-1293-1

[59] Y. Suetomi, S. Takahashi, T. Takao, H. Maeda , and Y. Yanagisawa, A novel winding method for a no-insulation layer-wound REBCO coil to provide a short

magnetic field delay and self-protect characteristics, Supercond. Sci. Technol. 32 (2019) 045003. https://doi.org/10.1088/1361-6668/ab016e

[60] E. Berrospe-Juarez, V. M. R. Zermeno, F. Trillaud, A. V. Gavrilin, F. Grilli, D. V. Abraimov, D. K. Hilton and H. W. Weijers, Estimation of losses in the (RE) BCO two-coil insert of the NHMFL 32 T all-superconducting magnet, IEEE Trans. Appl. Supercond. 28 (2018) 1-5. https://doi.org/10.1109/TASC.2018.2791545

[61] J. Liu, Q. Wang, L. Qin, B. Zhou, K. Wang, Y. Wang, L. Wang, Z. Zhang, Y. Dai, H. Liu, and X. Hu, World record 32.35-tesla direct-current magnetic field generated with an all-superconducting magnet, Supercond. Sci. Technol. 33 (2020) 03LT01. https://doi.org/10.1088/1361-6668/ab714e

[62] D. C. Larbalestier, L. D. Cooley, M. O. Rikel, A. A. Polyanskii, J. Jiang, S. Patnaik, X. Y. Cai, D. M. Feldmann, A. Gurevich, A. A. Squitieri and M. T.Naus, Strongly linked current flow in polycrystalline forms of the superconductor MgB2, Nature 410 (2001) 186-189. https://doi.org/10.1038/35065559

[63] Y. E. Shujun H. Kumakura, The development of MgB2 superconducting wires fabricated with an internal Mg diffusion (IMD) process, Supercond. Sci. Technol. 29 (2016) 113004. https://doi.org/10.1088/0953-2048/29/11/113004

[64] S. X. Dou, S. Soltanian, J. Horvat, X. L. Wang, S. H. Zhou, M. Ionescu, H. K. Liu, P. Munroe and M. Tomsic, Enhancement of the critical current density and flux pinning of MgB2 superconductor by nanoparticle SiC doping, Appl. Phys. Lett 81 (2002) 3419-3421. https://doi.org/10.1063/1.1517398

[65] Y. Ma, X. Zhang, G. Nishijima, K. Watanabe, S. Awaji and X. Bai, Significantly enhanced critical current densities in MgB2 tapes made by a scaleable nanocarbon addition route, Appl. Phys. Lett. 88 (2006) 072502. https://doi.org/10.1063/1.2173635

[66] G. Z. Li, M. D. Sumption, J. B. Zwayer, M. A. Susner, M. A. Rindfleisch, C. J. Thong, M. J. Tomsic and E. W. Collings, Effects of carbon concentration and filament number on advanced internal Mg infiltration-processed MgB2 strands, Supercond. Sci. Technol. 26 (2013) 095007. https://doi.org/10.1088/0953-2048/26/9/095007

[67] D. Wang, D. Xu, X. Zhang, C. Yao, P. Yuan, Y. Ma, H. Oguro, S. Awaji and K. Watanabe, Uniform transport performance of a 100 m-class multifilament MgB2 wire fabricated by an internal Mg diffusion process, Supercond. Sci. Technol. 29 (2016) 065003. https://doi.org/10.1088/0953-2048/29/6/065003

[68] M. Tomsic, M. Rindfleisch, J. Yue, K. McFadden, D. Doll, J. Phillips, M. D. Sumption, M. Bhatia, S. Bohnenstiehl and E. W. Collings, Development of magnesium diboride (MgB2) wires and magnets using in situ strand fabrication method, Physica C 456 (2007) 203-208. https://doi.org/10.1016/j.physc.2007.01.009

[69] R. Flukiger, Advances in MgB2 conductors, Applied Superconductivity Conference 2014, Charlotte, USA.

[70] T. Katase, Y. Ishimaru, A. Tsukamoto, H. Hiramatsu, T. Kamiya, K. Tanabe and H. Hosono, 2011. Advantageous grain boundaries in iron pnictide superconductors, Nat. Commun. 2 (2011) 1-6. https://doi.org/10.1038/ncomms1419

[71] Y. Ma, Progress in wire fabrication of iron-based superconductors, Supercond. Sci. Technol. 25 (2012) 113001. https://doi.org/10.1088/0953-2048/25/11/113001

[72] J. D. Weiss, C. Tarantini, J. Jiang, F. Kametani, A. A. Polyanskii, D. C. Larbalestier and E. E. Hellstrom, High intergrain critical current density in fine-grain (Ba0.6K0.4) Fe2As2 wires and bulks, Nature Mater. 11 (2012) 682-85. https://doi.org/10.1038/nmat3333

[73] A. Malagoli, E. Wiesenmayer, S. Marchner, D. Johrendt, A. Genovese and M. Putti, Role of heat and mechanical treatments in the fabrication of superconducting Ba0.6K0.4Fe2As2 ex situ powder-in-tube tapes, Supercond. Sci. Technol. 28 (2015) 095015. https://doi.org/10.1088/0953-2048/28/9/095015

[74] X. Zhang, C. Yao, H. Lin, Y. Cai, Z. Chen, J. Li, C. Dong, Q. Zhang, D. Wang, Y. Ma and H. Oguro, Realization of practical level current densities in Sr0.6K0.4Fe2As2 tape conductors for high-field applications, Appl. Phys. Lett. 104 (2014) 202601. https://doi.org/10.1063/1.4879557

[75] Z. Gao, K. Togano, Y. Zhang, A. Matsumoto, A. Kikuchi and H. Kumakura, High transport Jc in stainless steel/Ag-Sn double sheathed Ba122 tapes, Supercond. Sci. Technol. 30 (2017) 095012. https://doi.org/10.1088/1361-6668/aa7bb9

[76] B. Shabbir, H. Huang, C. Yao, Y. Ma, S. Dou, T. H. Johansen, H. Hosono and X. Wang, Evidence for superior current carrying capability of iron pnictide tapes under hydrostatic pressure, Phys. Rev. Mater. 1 (2017) 044805. https://doi.org/10.1103/PhysRevMaterials.1.044805

[77] H. Hosono, A. Yamamoto, H. Hiramatsu and Y. Ma, Recent advances in iron-based superconductors toward applications. Mater. Today 21 (2018) 278-302. https://doi.org/10.1016/j.mattod.2017.09.006

[78] C. Yao and Y. Ma, Recent breakthrough development in iron-based superconducting wires for practical applications. Supercond. Sci. Technol. 32 (2019) 023002. https://doi.org/10.1088/1361-6668/aaf351

[79] H. Huang, C. Yao, C. Dong, X. Zhang, D. Wang, Z. Cheng, J. Li, S. Awaji, H. Wen and Y. Ma, High transport current superconductivity in powder-in-tube Ba0.6K0.4Fe2As2 tapes at 27 T, Supercond. Sci. Technol. 31 (2017) 015017. https://doi.org/10.1088/1361-6668/aa9912

[80] S. Pyon, D. Miyawaki, T. Tamegai, S. Awaji, H. Kito, S. Ishida and Y. Yoshida, Enhancement of critical current density in (Ba, Na) Fe2As2 round wires using high-pressure sintering, Supercond. Sci. Technol. 33 (2020) 065001. https://doi.org/10.1088/1361-6668/ab804c

[81] S. Liu, C. Yao, H. Huang, C. Dong, W. Guo, Z. Cheng, Y. Zhu, S. Awaji and Y. Ma, High-performance Ba1−xKxFe2As2 superconducting tapes with grain texture engineered via a scalable fabrication. Sci. China Mater. 64 (2021) 2530-2540. https://doi.org/10.1007/s40843-020-1643-1

[82] G. Sylva, A. Augieri, A. Mancini, A. Rufoloni, A. Vannozzi, G. Celentano, E. Bellingeri, C. Ferdeghini, M. Putti and V. Braccini, Fe (Se, Te) coated conductors deposited on simple rolling-assisted biaxially textured substrate templates. Supercond. Sci. Technol. 32 (2019) 084006. https://doi.org/10.1088/1361-6668/ab0e98

[83] I. Pallecchi, C. Tarantini, J. Hänisch and A. Yamamoto, Preface to the special issue 'Focus on 10 Years of Iron-Based Superconductors', Supercond. Sci. Technol. 33 (2020) 090301. https://doi.org/10.1088/1361-6668/ab9ad2

[84] J. Guo, S. Jin, G. Wang, S. Wang, K. Zhu, T. Zhou, M. He and X. Chen, Superconductivity in the iron selenide KxFe2Se2 (0≤x≤1.0), Phys. Rev. B 82 (2010) 180520. https://doi.org/10.1103/PhysRevB.82.180520

[85] X. F. Lu, N. Z. Wang, H. Wu, Y. P. Wu, D. Zhao, X. Z. Zeng, X. G. Luo, T. Wu, W. Bao, G. H. Zhang and F. Q. Huang, Coexistence of superconductivity and antiferromagnetism in (Li0.8Fe0.2)OHFeSe, Nat. Mater. 14 (2015) 325-329. https://doi.org/10.1038/nmat4155

[86] C. Yao, Y. Ma, X. Zhang, D. Wang, C. Wang, H. Lin and Q. Zhang, Fabrication and transport properties of Sr0.6K0.4Fe2As2 multifilamentary superconducting wires, Appl. Phys. Lett. 102 (2013) 082602. https://doi.org/10.1063/1.4794059

[87] X. Zhang, H. Oguro, C. Yao, C. Dong, Z. Xu, D. Wang, S. Awaji, K. Watanabe and Y. Ma, Superconducting properties of 100-m class Sr0.6K0.4Fe2As2 tape and pancake coils, IEEE Trans. Appl. Supercond. 27 (2017) 1-5. https://doi.org/10.1109/TASC.2017.2650408

[88] D. Wang, Z. Zhang, X. Zhang, D. Jiang, C. Dong, H. Huang, W. Chen, Q. Xu and Y. Ma, First performance test of a 30 mm iron-based superconductor single pancake coil under a 24 T background field, Supercond. Sci. Technol. 32 (2019) 04LT01. https://doi.org/10.1088/1361-6668/ab09a4

[89] X. Qian, S. Jiang, H. Ding, P. Huang, G. Zou, D. Jiang, X. Zhang, Y. Ma and W. Chen, Performance testing of the iron-based superconductor inserted coils under high magnetic field, Physica C: Superconductivity and its Applications, 580 (2021) 1353787. https://doi.org/10.1016/j.physc.2020.1353787

[90] Z. Zhang, D. Wang, S. Wei, Y. Wang, C. Wang, Z. Zhang, H. Yao, X. Zhang, F. Liu, H. Liu and Y. Ma, First performance test of the iron-based superconducting racetrack coils at 10 T, Supercond. Sci. Technol. 34 (2021) 035021. https://doi.org/10.1088/1361-6668/abb11b

[91] The CEPC Study Group, CEPC Conceptual Design Report, Vol 1 - Accelerator, IHEP-CEPC-DR-2018-01, IHEP-AC-2018-01, August 2018.

[92] Z. Gao, L. Wang, Y. Qi, D. Wang, X. Zhang and Y. Ma. Preparation of LaFeAsO0.9F0.1 wires by the powder-in-tube method, Supercond. Sci. Technol. 21 (2008) 105024. https://doi.org/10.1088/0953-2048/21/10/105024

[93] Y. Qi, X. Zhang, Z. Gao, Z. Zhang, L. Wang, D. Wang and Y. Ma, Superconductivity of powder-in-tube Sr0.6K0.4Fe2As2 wires, Phys. C: Supercond. 469 (2009) 717-720. https://doi.org/10.1016/j.physc.2009.03.008

[94] H. Hilgenkamp and J. Mannhart, Grain boundaries in high-Tc superconductors, Rev. Mod. Phys. 74 (2002) 485. https://doi.org/10.1103/RevModPhys.74.485

[95] C. Yao, H. Lin, Q. Zhang, X. Zhang, D. Wang, C. Dong, Y. Ma, S. Awaji and K. Watanabe, Critical current density and microstructure of iron sheathed multifilamentary Sr1−xKxFe2As2/Ag composite conductors, J. Appl. Phys. 118 (2015) 203909. https://doi.org/10.1063/1.4936370

[96] X. Zhang, C. Yao, H. Lin, Y. Cai, Z. Chen, J. Li, C. Dong, Q. Zhang, D. Wang, Y. Ma and H. Oguro, Realization of practical level current densities in Sr0.6K0.4Fe2As2 tape conductors for high-field applications, Appl. Phys. Lett. 104 (2014) 202601. https://doi.org/10.1063/1.4879557

[97] C. Yao, H. Lin, Q. Zhang, X. Zhang, D. Wang, C. Dong, Y. Ma, S. Awaji and K. Watanabe, Critical current density and microstructure of iron sheathed multifilamentary Sr1−xKxFe2As2/Ag composite conductors, J. Appl. Phys. 118 (2015) 203909. https://doi.org/10.1063/1.4936370

[98] A. Gurevich S. Patnaik, V. Braccini, K. H. Kim, C. Mielke, X. Song, L. D. Cooley, S. D. Bu, D. M. Kim, J. H. Choi and L. J. Belenky, Very high upper critical fields in MgB2 produced by selective tuning of impurity scattering, Supercond. Sci. Technol. 17 (2004) 278. https://doi.org/10.1088/0953-2048/17/2/008

[99] J. Jaroszynski, F. Hunte, L. Balicas, Y. J. Jo, I. Raičević, A. Gurevich, D. C. Larbalestier, F. F. Balakirev, L. Fang, P. Cheng and Y. Jia, Upper critical fields and thermally-activated transport of NdFeAsO 0.7 F 0.3 single crystal, Phys. Rev. B 78 (2008) 174523. https://doi.org/10.1103/PhysRevB.78.174523

[100] C. Tarantini, A. Gurevich, J. Jaroszynski, F. Balakirev, E. Bellingeri, I. Pallecchi, C. Ferdeghini, B. Shen, H. H. Wen and D. C. Larbalestier, Significant enhancement of upper critical fields by doping and strain in iron-based superconductors, Physical Review B 84 (2011) 184522. https://doi.org/10.1103/PhysRevB.84.184522

[101] Y. Ando, G. S. Boebinger, A. Passner, L. F. Schneemeyer, T. Kimura, M. Okuya S. Watauchi, J. Shimoyama, K. Kishio, K. Tamasaku and N. Ichikawa, Resistive upper critical fields and irreversibility lines of optimally doped high-Tc cuprates. Phys. Rev. B, 60 (1999) 12475. https://doi.org/10.1103/PhysRevB.60.12475

[102] I. Matsubara, H. Tanigawa, T. Ogura, H. Yamashita. M. Kinoshita and T. Kawai, Upper critical field and anisotropy of the high-Tc Bi2Sr2Ca2Cu3O x phase, Phy. Rev. B 45 (1992) 7414.

Materials Research Forum LLC
https://doi.org/10.21741/9781644902110-13

[103] L. Quettier, , G. Aubert, , J. Belorgey, C. Berriaud, A. Bourquard, Ph. Bredy, O. Dubois, G. Gilgrass, F.P Juster, H. Lannou, F. Molinié, M. Nusbaum, F. Nunio, A. Payn, T. Schild, M. Schweitzer, L. Scola, A. Sinanna, V. Stepanov and P. Vedrine, Iseult/INUMAC Whole Body 11.7 T MRI Magnet, IEEE Trans. Appl. Supercond. 25 (2015) 4301404 https://doi.org/10.1109/TASC.2014.2369233

[104] P. Tixador, C. E. Bruzek, B. Vincent, A. Malgoli and X. Chaud, Mechanically reinforced Bi-2212 strand, IEEE Trans. Appl. Supercond. 25 (2014) 1-4. https://doi.org/10.1109/TASC.2014.2373642

[105] T. Shen, P. Li, J. Jiang, L. Cooley, J. Tomopkins, D. McRae and R. Walsh, High strength kiloampere Bi2Sr2CaCu2Ox cables for high-field magnet applications, Supercond. Sci. Technol. 28 (2015) 065002. https://doi.org/10.1088/0953-2048/28/6/065002

[106] C. Yao, H. Lin, Q. Zhang, X. Zhang, D. Wang, C. Dong, Y. Ma, S. Awaji and K. Watanabe, Critical current density and microstructure of iron sheathed multifilamentary Sr1-xKxFe2As2/Ag composite conductors, J. Appl. Phys. 118 (2015) 203909. https://doi.org/10.1063/1.4936370

Keyword Index

About the Editors

Dr. Inamuddin is working as Assistant Professor at the Department of Applied Chemistry, Aligarh Muslim University, Aligarh, India. He obtained Master of Science degree in Organic Chemistry from Chaudhary Charan Singh (CCS) University, Meerut, India, in 2002. He received his Master of Philosophy and Doctor of Philosophy degrees in Applied Chemistry from Aligarh Muslim University (AMU), India, in 2004 and 2007, respectively. He has extensive research experience in multidisciplinary fields of Analytical Chemistry, Materials Chemistry, and Electrochemistry and, more specifically, Renewable Energy and Environment. He has worked on different research projects as project fellow and senior research fellow funded by University Grants Commission (UGC), Government of India, and Council of Scientific and Industrial Research (CSIR), Government of India. He has received Fast Track Young Scientist Award from the Department of Science and Technology, India, to work in the area of bending actuators and artificial muscles. He has also received the Sir Syed Young Researcher of the Year Award 2020 from Aligarh Muslim University. He has completed four major research projects sanctioned by University Grant Commission, Department of Science and Technology, Council of Scientific and Industrial Research, and Council of Science and Technology, India. He has published 205 research articles in international journals of repute and nineteen book chapters in knowledge-based book editions published by renowned international publishers. He has published 155 edited books with Springer (U.K.), Elsevier, Nova Science Publishers, Inc. (U.S.A.), CRC Press Taylor & Francis Asia Pacific, Trans Tech Publications Ltd. (Switzerland), IntechOpen Limited (U.K.), Wiley-Scrivener, (U.S.A.) and Materials Research Forum LLC (U.S.A). He is a member of various journals' editorial boards. He is also serving as Associate Editor for journals (Environmental Chemistry Letter, Applied Water Science and Euro-Mediterranean Journal for Environmental Integration, Springer-Nature), Frontiers Section Editor (Current Analytical Chemistry, Bentham Science Publishers), Editorial Board Member (Scientific Reports-Nature), Editor (Eurasian Journal of Analytical Chemistry), and Review Editor (Frontiers in Chemistry, Frontiers, U.K.). He is also guest-editing various special thematic special issues to the journals of Elsevier, Bentham Science Publishers, and John Wiley & Sons, Inc. He has attended as well as chaired sessions in various international and national conferences. He has worked as a Postdoctoral Fellow, leading a research team at the Creative Research Initiative Center for Bio-Artificial Muscle, Hanyang University, South Korea, in the field of renewable energy, especially biofuel cells. He has also worked as a Postdoctoral Fellow at the Center of Research Excellence in Renewable Energy, King Fahd University of Petroleum and Minerals, Saudi Arabia, in the field of polymer electrolyte membrane fuel cells and computational fluid dynamics of polymer electrolyte membrane fuel cells. He is a life member of the Journal of the Indian Chemical Society. His research interest includes ion exchange materials, a sensor for heavy metal ions, biofuel cells, supercapacitors and bending actuators.

Tariq Altalhi, PhD, is working as Associate Professor in the Department of Chemistry at Taif University, Saudi Arabia. He received his doctorate degree from University of Adelaide, Australia in the year 2014 with Dean's Commendation for Doctoral Thesis Excellence. He has worked as head of Chemistry Department at Taif university and Vice Dean of Science college. In 2015, one of his works was nominated for Green Tech awards from Germany, Europe's largest environmental and business prize, amongst top 10 entries. He has co-edited various scientific books. His group is involved in fundamental multidisciplinary research in nanomaterial synthesis and engineering, characterization, and their application in molecular separation, desalination, membrane systems, drug delivery, and biosensing. In addition, he has established key contacts with major industries in Kingdom of Saudi Arabia.

Dr. Vikas Gupta, is a Professor and Dean, Faculty of Science at Motherhood University, Roorkee, Uttrakhand, India. He has published several research papers of national and international repute. He has contributed immensely in advanced areas of chemistry like polymer science, environmental chemistry and chromatography. His contribution to modern areas of chemistry is well recognized and widely cited. He has contributed as an investigator in an Armament Research Board of DRDO project at the College of Engineering and Technology, Moradabad. He has wide experience in teaching undergraduate and postgraduate students of science and engineering. Dr. Gupta is a lifetime member of the Indian Science Congress. He has supervised several under his guidance. He has organized DRDO-sponsored National Conference as well as various webinars in the field of science.

Dr. Mohammad Luqman has 12+ years of post-PhD experience in Teaching, Research, and Administration. Currently, he is serving as an Assistant Professor of Chemical Engineering in Taibah University, Saudi Arabia. Before joining here, he served as an Assistant Professor in College of Applied Science at A'Sharqiyah University, Oman, and in College of Engineering at King Saud University, Saudi Arabia. He served as a Research Engineer in SAMSUNG Cheil Industries, South Korea. Moreover, he served as a post-doctoral fellow at Artificial Muscle Research Center, Konkuk University, South Korea, in the field of Ionic Polymer Metal Composites for the development of Artificial Muscles, Robotic Actuators and Dynamic Sensors. He earned his PhD degree in the field of Ionomers (Ion-containing Polymers), from Chosun University, South Korea. He successfully served as an Editor to three books, published by world renowned publishers. He published numerous high-quality papers, and book chapters. He is serving as an

Editor and editorial/review board members to many International SCI and Non-SCI journals. He has attracted a few important research grants from industry and academia. His research interests include but not limited to Development of Ionomer/Polyelectrolyte/non-ionic Polymer Nanocomposites/Blends for Smart and Industrial/Engineering Applications.

www.ingramcontent.com/pod-product-compliance
Lightning Source LLC
Chambersburg PA
CBHW071340210326
41597CB00015B/1519